Heterogeneous Catalysis

Heterogeneous Catalysis

Fundamentals and Applications

Julian R.H. Ross

ELSEVIER

AMSTERDAM • BOSTON • HEIDELBERG • LONDON • NEW YORK • OXFORD
PARIS • SAN DIEGOSAN FRANCISCO • SINGAPORE • SYDNEY • TOKYO

Elsevier
Radarweg 29, PO Box 211, 1000 AE Amsterdam, The Netherlands
The Boulevard, Langford Lane, Kidlington, Oxford OX5 1GB, UK

British Library Cataloguing in Publication Data
A catalogue record for this book is available from the British Library

Library of Congress Cataloging-in-Publication Data
A catalog record for this book is available from the Library of Congress

ISBN: 978-0-444-63840-3

For information on all Elsevier publications
visit our website at www.elsevierdirect.com

11 12 10 9 8 7 6 5 4 3 2 1

Contents

Preface

A former colleague used to say that to sound convincing, one should only make one excuse at a time. Nevertheless, I want to give several excuses for having written yet another textbook on catalysis.

My first excuse is that, despite the large number of books which cover the field of catalysis, I have never found one that does exactly what I want: gives a general introduction to the main aspects of the subject and then gives treatments of the most important topics without giving too many details. There are many scholarly texts (to some of which I refer quite frequently in this book) which treat the subject in great detail and which give exhaustive reviews of the relevant literature but which are, in most aspects, too advanced for general undergraduate use. There are others which are more student friendly and approach the subject in a more readable fashion but do not cover all the important aspects of the subject.

My second excuse is that I want to try to do something that is slightly new. Many years ago, I was introduced to the concept of Citation Indexing by a remark from a former colleague: "The future of the use of the chemical literature lies in citation tools such as the Science Citation Index." I then followed with great interest over the years, and was greatly inspired by, a series of essays in "Current Contents" by Dr. Eugene Garfield of the Institute of Scientific Information devoted to the use of such tools. In essence, citation indexing allows one to move *forward* in the literature – rather than *back* in the way that we had been trained to do using "Chemical Abstracts" and similar indexing tools. What I am trying to do in this book is to encourage the student to use such tools to move forward from a key reference on a specific topic so that he or she can find out what is happening in that field.[1] I encourage the student when embarking on such searches to use wherever possible "secondary sources" (in other words, review articles and such publications) rather than getting tied down in too much of the detail found in research papers. Further, I want to encourage the student not to try to read every detail of an article (one can always return to it later if ones interests change) but to dig out the most pertinent details for the purpose in hand.

[1] Another great advantage of the use of Scopus or a similar indexing tool is that it ensures that the text becomes less obsolete with time than it might otherwise be: the student is always able to find what is happening at that particular moment in a particular subject. The main problem is that there may be new "hot subjects" and, for these, there will most probably be a need for updates of the main text to give pointers to these topics.

My third excuse is that I have been fortunate in my working life at three different universities in three different countries (having studied in yet another establishment) to have encountered a wide range of different scientists working on different aspects of catalysis. On top of that, I have been privileged to have been Editor of *Catalysis Today* (published by Elsevier Science Publishers) for more than 20 years and so I have had the chance of widening my knowledge (still unfortunately quite superficial in some fields) of a large number of sub-divisions of the subject. As a result of this work, which has coincided with the vast expansion of the use of electronic communication, I have also become more and more aware of the advantages of the use of electronic literature resources in day-to-day scientific life. Hence, I have wanted to put that knowledge to good use.

My fourth excuse is that I have had very positive experience of the use of project-based learning of the type required here. I used this method particularly effectively during my time at the University of Twente (1982–1991) when, following a short course of lectures on more fundamental aspects of applied catalysis, I asked each student to perform, under my guidance, a detailed literature search on a specific subject and then to present a seminar on the results to the whole class. The results were very good and several of the projects even led to new research themes in my group.

My final excuse is that I believe that catalysis is a subject which should be taught even more widely than it is. Catalysis is a topic that impinges on so many different aspects of chemistry and chemical engineering that every chemist and chemical engineer should be taught the fundamentals of the subject; further, they should also be able to use the appropriate scientific and technical literature to enable them, if necessary, later to become more expert on specific aspects.

The book before you (either in paper or in electronic form) has emerged gradually during the writing process from being purely another textbook on the subject (including many of the pet themes of the author – although some of those of this author remain in this text) to being an interactive and open-ended document that allows the student to pursue the topics of most interest to him or her. No attempt is made to give too much detail and the student is encouraged to aim to gain an understanding of the basics rather than to delve in too deeply, at least at the time of first reading.[2] The text includes a series of boxes and tasks (which include subsidiary information and suggestions for further literature work) and the student can either omit these on the first reading or make a selection of which to read in more detail. The book can be used either for private study or as a class textbook under the guidance of a lecturer/instructor. In the latter case, the lecturer/instructor can guide the student in the use of the literature and steer

[2] You can always go back to a particular subject later if necessary. Remember, it is not what you know but knowing where to find the relevant information that is likely to be of greatest importance in your later career.

them in particular directions; he/she can, if wished, also use the tasks set in the text for grading purposes.

Despite my remarks above about pet themes, it will be clear to the reader that I have allowed myself to use many examples from my personal experience. Further, I have used many key references gleaned from Elsevier journals, particularly *Catalysis Today, Applied Catalysis* and *Journal of Catalysis*. While I could also have used papers published by other publishers, I have tried to restrict myself to papers which can easily be accessed via Science Direct. This does not mean that you should always restrict yourself to using Scopus or Science Direct. If you have access to the Web of Science or Scifinder or other electronic resources, by all means use these for your work and ignore the Elsevier material if you so wish. The end result should be similar: having studied the course outlined in this book, you should have a greater awareness of the importance of catalysis and of many of its applications.

Appendix: Methodology to be Used

As we will use citation methods as an integral part of this text, it is appropriate now to say a little more about such methods as a means of researching the scientific literature.[3] When a newly published paper is indexed, the papers that it cites are linked to the new paper and full details of the authors and the abstract of the paper are also stored. Links are also provided to full-text versions of the papers, access to these being limited by the subscription under which the search is being carried out. Most importantly, because of the citation listing, it is possible to work forward in the literature from a source article. Suppose that one is aware of a very important source article (perhaps a comprehensive review of a particular subject) published 10 years ago and one wants to find out what has been added to the literature since, one looks up the original article and gets a list of the papers that have cited it. Depending on the number of "cites", one can then either look up each of those papers (or, at least, read the abstracts) or refine the list further by excluding titles that are peripheral to the subject of interest before going further. [4] (One can also find the most important of the citing articles by ranking them by number of citations.)

[3] It should be noted that citation analysis has been adopted as a method of assessing scientific output as it is very easy to find the number of times that a particular author has been cited and to determination of the so-called Hirsch Index (h-index), the number of papers, h, which have been cited h or more times. It should be remembered, but often isn't, that both these quantities are very dependent on the field in which a particular scientist is working, on the popularity of that field, and on factors such a whether or not the author has written significant reviews or methodological papers. (Someone working in a relatively obscure field and writing one or two very significant papers a year is much less likely to be cited than someone working on a "hot" topic and publishing prolifically.) The methodology is therefore wide open to abuse unless used very carefully.

[4] Current awareness of a specific topic can also be achieved by storing a number of key references on that subject and requesting electronic alerts to any papers citing those references.

We will use Scopus or Web of Science to aid us in the literature studies which are included as "tasks" in this text. Throughout the book, a number of topics are suggested for further study and details of authors publishing in that area or of several important papers or reviews in that area of research are given. The student is then encouraged to follow up on each topic using Scopus or Web of Science (and the links available from them) to select the most recent and significant publications, either reviews or full scientific articles, on the subject. Each student will follow up the subject in a different way, depending on his or her own interests and specific requirements: one will be more interested in catalyst materials, another in process aspects of a reaction, another in economic assessment of new processes, etc. When the text is used in conjunction with a course of lectures, this approach also allows the instructor to have an input in the choice of topics or to suggest new ones relevant to the research interests of the department in which the course is being given. The instructor can also, if required, use these studies as the basis for assessment rather than setting formal examinations. In my view, this is the preferable approach since the methodology suggested allows the student to exhibit his or her own initiative and understanding of a subject much better than would be possible under examination conditions.

Acknowledgements

I would like to thank the many research students and post-docs who have worked with over the years, too many to mention by name, for their hard work and enthusiasm. Through them, I have been able to widen my knowledge of the subject of catalysis, a subject that has enthralled me since my own student days. I also wish to thank all my various colleagues in three different universities in which I have worked (Bradford, UK, 1966–1982; Twente, NL, 1982–1991 and Limerick, Ireland, 1991–present) for their contributions, in one way or another, over my working life.

But most importantly, I want to thank my wife, Anne, for her constant support; without her help and understanding, this book would never have appeared.

Julian Ross
January 2011

CHAPTER 1

Heterogeneous Catalysis — Chemistry in Two Dimensions

Chapter Outline

1.1 Introduction

I assume that you are a chemist, or at least that you have enough understanding of chemistry to be able to understand its language and short-hand, and that you understand chemical equations such as[1]:

$$CO + H_2O = CO_2 + H_2 \tag{1.1}$$

or even abstract ones such as:

$$A + B = C + D \tag{1.2}$$

You will recognize that the equals sign (=) means that the equation is balanced and that the reaction is (mostly) at equilibrium; in some cases, there is a reversible arrow ($\rightleftarrows$) indicating that both the forward and reverse reactions occur. You will probably also realize

[1] This is the so-called water–gas shift reaction. This reaction is of importance in determining the equilibrium composition in many processes such as steam reforming of methane or methanol synthesis.

Heterogeneous Catalysis. DOI: 10.1016/B978-0-444-53363-0.10001-5

that the equations could have associated with them the enthalpies of reaction (for Eq. 1.1, $\Delta H^o = -40.6\ kJ\ mol^{-1}$).[2] Further, you may recognize that if the equals sign is substituted by an arrow ($\rightarrow$), we are more likely to be dealing with a reaction controlled by kinetics than by thermodynamics. However, when we learn organic chemistry or inorganic chemistry, we sometimes forget these niceties and just worry about what products are formed when we add two chemicals together. And when we read an equation such as:

$$C_2H_4 + H_2 \xrightarrow{Ni} C_2H_6 \qquad (1.3)$$

we accept that the reaction is catalysed by Ni – but often without asking why or how. Glance at your organic textbook (at least, if it is the type that I used when I was a student) and you will see many such "qualified" arrows, often without any explanation or rationale. The catalyst is a "Black Box". My aim in writing this textbook, from which I hope you will be able to benefit significantly, with or without the help of a lecturer or instructor, is that you should, when you have finished studying it, be able to understand all the parameters associated with such equations and have a much deeper understanding of what a catalyst is, how it is made and applied and how it works (or doesn't work); and you should also be in a position to study the literature on in a critical way in order to gain information on any reaction about which you may have an interest. We will start in this chapter with a short exploration of the history of catalysis, with particular reference to heterogeneous catalysis and will then move on to cover some of the fundamental aspects of catalysis (catalyst preparation, characterization, experimental methods of studying catalysis, the kinetics of catalytic reactions, etc.), before moving on to examine a number of catalytic reactions of current importance. The approach, as discussed in the Preface, will be based largely on the use of literature accessible through the internet, this methodology being particularly important in the chapters dealing with modern catalytic processes. You, the reader, are encouraged throughout to read widely round the subject and to explore the most up-to-date literature on each subject considered.

1.2 Historical Background to Catalysis

Various examples of catalysis, both heterogeneous and homogeneous, have been known for many centuries. Perhaps the earliest example of catalysis was the use of yeast, a material which

[2] The value given should be the standard enthalpy, $\Delta H^o{}_{298}$ (i.e., the heat of reaction for the reactants and products in their standard states at a reaction temperature of 298 K) but it may also be the enthalpy at the temperature of reaction given. In the Russian literature, the enthalpy is often included in the equation; in that case, the enthalpy of reaction appears on the right hand-side of the equation as a positive quantity if the reaction gives out heat (i.e. it is exothermic) or with a negative sign if the reaction requires heat (i.e. it is endothermic).

contains an enzyme that brings about the fermentation of the sugar contained in biological materials such as grain or grapes to give ethyl alcohol. Fermentation has been known for some 8000 years: beer was first produced in Ancient Egypt and Mesopotamia (today's Iraq) while the earliest wine production was in Georgia and Iran.

BOX 1.1 Homogeneous versus Heterogeneous Catalysis

A homogeneous catalytic process is one in which the catalyst is in the same phase as the reactants and products. A simple example is the use of an acid to catalyse the hydrolysis of an ester (or the reverse esterification process):

$$RCOOR' + H_2O \xrightarrow{H^+} RCOOH + R'\,OH$$

The rate of reaction in the presence of H^+ ions is very much higher than in their absence and yet the H^+ ion does not enter into the stoichiometric equation and is therefore the catalyst. The proton of the acid reacts with the ester as follows:

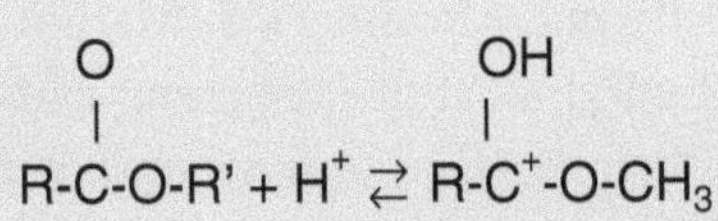

and the ion that is formed is much more susceptible to nucleophilic attack by the water molecule than is the original ester molecule:

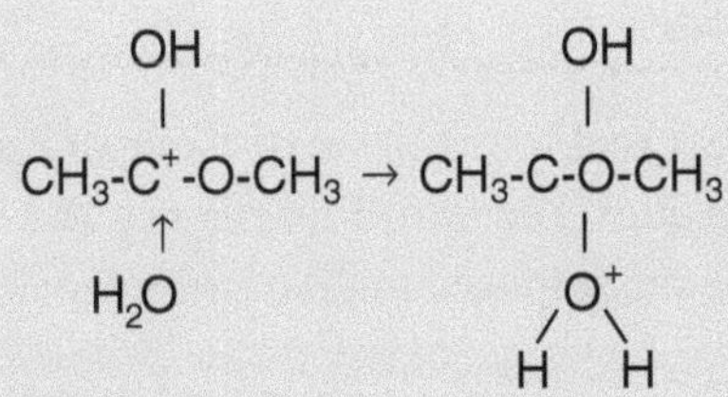

The compound formed then rearranges by a redistribution of electrons:

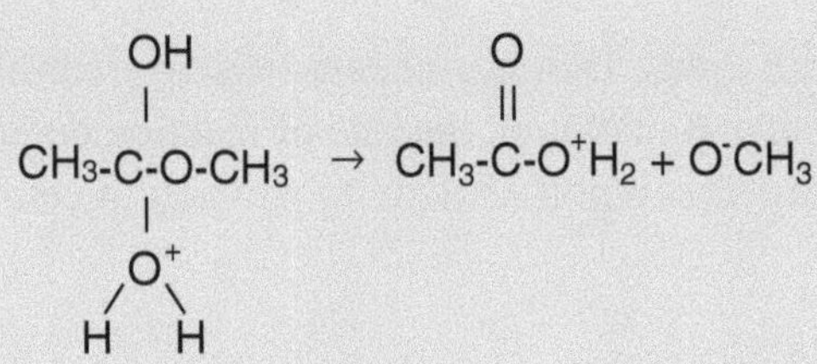

Finally, the products are formed, liberating the proton once more.

Other homogeneously catalysed reactions include those in which an inorganic complex is involved. An example is the Wilkinson catalyst:

$$RhCl(PPh_3)_3 (Ph = phenyl)$$

(Continued)

BOX 1.1 Homogeneous versus Heterogeneous Catalysis—*Cont'd*

which can be used for the hydrogenation of double bonds and for other reactions in the liquid phase. (http://en.wikipedia.org/wiki/Wilkinson's_catalyst).

In contrast, a heterogeneous catalyst is one which exists in a different phase to that of the reactants; the catalyst is usually a solid and the reactants are either gases or liquids. Examples of such catalysts are give in the main text and the mechanism of heterogenous catalysis will be discussed in detail in Chapter 6. It is worth noting that the reactions discussed above, the hydrolysis of an ester or the reverse reaction (esterification of an acid) can be carried out using a solid acid catalyst.

The earliest work on heterogeneous catalysis, as we know it, may have been done by the alchemists in their search for a route to gold from base metals. However, the first formal scientific reports of the use of heterogeneous catalysts were written independently in the year 1800 by Joseph Priestly and Martinus van Marum, both of whom reported work on the dehydrogenation of ethyl alcohol over metal catalysts. Surprisingly, at least to the modern scientist, neither of them appears to have recognized that the metals were acting as catalysts, seemingly thinking that the metals just supplied heat for the reaction.

1.2.1 Ammonia Decomposition

What seems to have been the first real recognition of the operation of a heterogeneous catalyst was made in 1813 by Louis Jacques Thénard, Professor at the École Polytechnique in Paris, who reported that ammonia was decomposed to give hydrogen and nitrogen when it was passed over red-hot metals:

$$NH_3 \rightarrow N_2 + 3H_2$$

Ten years later, working with Pierre Dulong, Thénard discovered that the reaction occurred over iron, copper, silver, gold and platinum, the rate of reaction decreasing in the order given. This appears to have been the first report of activity patterns in catalysis.

1.2.2 Catalytic Oxidation

Shortly after the initial observation had been reported by Thénard, one of the most important early experiments in the development of heterogeneous catalysis, as we know it, was carried out in the laboratories of the Royal Institution in London in 1817: Humphrey Davy, assisted by a young Michael Faraday, found that a heated platinum wire could bring about the combination of air and coal gas (largely CO and H_2) without the action of a flame. This was the first reported example of catalytic oxidation. Davy reproduced the result with palladium but failed to do so with copper, silver, iron, gold and zinc. The great importance of the surface area of the platinum was demonstrated in 1820 at University College, Cork by Davy's cousin, Edmond Davy, who showed that finely divided platinum could oxidize alcohol at room temperature. Edmond Davy's technique for making high area platinum was improved in 1823 by J.W. Dobreiner, working at the University of Jena, who produced a spongy material that caused the combination of hydrogen and oxygen at room temperature by hydrogen reduction of finely divided platinum. Later that year, the same reaction was shown by Dulong and Thénard to occur on palladium and iridium at around room temperature and on cobalt, nickel, rhodium, silver and gold when the temperature was increased.

BOX 1.2 The Miners' Lamp

The fact that the platinum brought about the reaction smoothly, without explosion, and yet with the evolution of heat which caused the wire to glow red-hot, was the basis for the development of the miners' safety lamp. A description of the work carried out by Davy is given by L.B. Hunt in *Platinum Metals Rev.*, 1979, **23**, (l), 29–31(http://www.platinummetalsreview.com/dynamic/article/view/pmr-v23-i1-029-031) and the early work on catalysis is summarized by A.J.B. Robertson in *Platinum Metals Rev.*, 1975, **19**, (2), 64–69 (http://www.platinummetalsreview.com/dynamic/article/view/pmr-v19-i2-064-069). Note that in practice the miners' lamp did not contain Pt but only an iron gauze. This gauze rusted in high water vapour pressures and so the lamp was in fact ineffective.

TASK 1.1 Sir Humphrey Davy

Using a search engine such as Google, do a search on Sir Humphrey Davy and read about some of his work at the Royal Institution. Pay particular attention to his work on the miners' lamp (Box 1.2) and to parallel work by Clanny and Stephenson. Examine the material available on the web in relation to the miner's lamp and see to what extent the lamp performed a catalytic reaction and to what extent the gauze of the lamp just provided a thermally conducting barrier between the flame and any gas to which it was exposed. Find out also about the use of the lamp as a gas detector.

Examine some of the other achievements of Davy and his relationships with other scientists working in the same period.

1.2.3 Berzelius and the Concept of Catalysis

As well as being a series of examples illustrating the development of catalytic oxidation, the work outlined above shows the way in which new ideas developed internationally, even in the nineteenth century. There was at that time much discussion on the way in which catalysts function. In an article summarizing some of the results outlined above presented to the Stockholm Academy of Sciences in 1835, J.J. Berzelius first introduced the term "catalysis". This word comes from the Greek word *loosen*. Berzelius proposed that there was a *catalytic force* and that *catalysis* occurred because of the decomposition of bodies by this force. The ideas behind the concept of what we now know as *adsorption* was introduced a year earlier in 1834 by Michael Faraday who had been working on electrolysis and had observed a spontaneous combustion of hydrogen and oxygen occurring on platinum electrodes. Faraday proposed that the two reacting gases were *condensed* on the surface of the platinum and that "the approximation of the particles to those of the metals may be very great". Faraday's proposition stressed the importance of the surface of the catalyst and also hinted at something that is now generally accepted: that the geometry of the interaction at the interface is also critical.

TASK 1.2 Michael Faraday

Having worked as Davy's assistant, Michael Faraday later became Director of the Royal Institution. During a very long career, he carried out research on a wide variety of topics and he is recognized as having been one of the most successful practical scientists of all time. Using a web-based search engine, examine some of the material available on his life and times, paying particular attention to his work on catalytic combustion.

1.2.4 The First Industrial Catalytic Processes

Examination of the interaction between molecules and surfaces ("chemisorption") had to wait until techniques were developed that allowed such studies to occur, these being dependent on the perfection of vacuum technologies, as will be discussed later (Section 1.2.7). However, a lack of a full understanding of the importance of chemisorption did not hinder the development of a number of important industrial processes. For example, the "Contact Process" for the production of sulfuric acid was patented by Peregrine Phillips, Junior, in 1831. Phillips brought about the oxidation of sulfur dioxide (SO_2) by drawing it with air through tubes of platinum (or similar material) packed with platinum wire or platinum in any finely divided state and heating them to a "strong yellow heat". (http://www.platinummetalsreview.com/dynamic/article/view/pmr-v19-i2-064-069). This reaction was not developed as an industrial process until 1875 when Messel introduced the industrial contact acid process; the process originally operated using a supported platinum catalyst but now uses a vanadia-containing material.

http://en.wikipedia.org/wiki/Sulfuric_acid#Manufacture
http://www.lenntech.com/Chemistry/Sulfuric-acid.htm
http://www.chemguide.co.uk/physical/equilibria/contact.html

The foundations for another important catalytic oxidation process, the oxidation of ammonia over platinum to give nitric oxide, a step in the formation of nitric acid and hence nitrate fertilizers, was reported by C.F. Kuhlmann in 1838.[3] Kuhlmann recognized that the process (using ammonia from coke-oven gas) would in due course become important even though at that time it was uncompetitive due to the cheap and readily available supplies of nitrogenous fertilizer in the form of Chile saltpeter. Towards the end of the nineteenth century, fears began to emerge that it would become impossible to feed the world's population using the available supply of this saltpeter. However, it was only with the threatened approach of war that scientists in various countries started to examine the oxidation of ammonia (then derived from coal gas) in detail. Hunt[3] describes the work carried out by Wilhelm Ostwald and his assistant Eberhard Brauer at the University of Leipzig in the first decade of the twentieth century. Starting by using platinized asbestos, they progressed to using a reactor containing a corrugated platinum strip. It fell to Karl Kaiser of the Technische Hochschule, Charlottenburg to develop in 1909 the Pt–Rh gauze-type catalysts of the type used until the present day.[4]

1.2.5 Ammonia Synthesis

It was also recognized at the turn of the century that the existing supplies of ammonia extracted made from coal gas would be inadequate as a basis for a fertilizer industry; for example, in his presidential address to the British association in 1898, Sir William Crookes appealed to the scientific community to develop a method of fixing atmospheric nitrogen to solve what was then termed the "Nitrogen Problem".

H.L. Le Chatelier, aware that the decomposition of ammonia should be reversible, did experiments in 1901 attempting to bring about the synthesis of ammonia from the elements. He carried out his experiments in a bomb at 200 atmospheres pressure and at a temperature of 600 °C using a reduced iron catalyst heated by a platinum spiral. However, he terminated his experiments after contamination of the reaction mixture with air led to an explosion. Le Chatelier is quoted as saying, near the end of his life, "I let the

[3] L. B. Hunt, *Platinum Metals* Rev., 1958,2,129
[4] http://www.platinummetalsreview.com/pdf/pmr-v2-i4-129-134.pdf

discovery of the ammonia synthesis slip through my hands. It was the greatest blunder of my scientific career".[5]

BOX 1.3 Ammonia Oxidation

It is well established that the Pt—Rh gauzes used for ammonia oxidation gradually become roughened and develop new structures during use, even though the chemical composition of the catalyst does not change appreciably. An article in Platinum Metals Review by F. Sperner and W. Hohmann (20 (1976) 12); http://www.platinummetalsreview.com/pdf/pmr-v20-i1-012-020.pdf) includes a series of electron micrographs of gauzes after use for some time. These demonstrate quite clearly that enormous physical change has occurred in the catalyst during use.

TASK 1.3 Fritz Haber

Fritz Haber, in common with Sir Humphrey Davy (Task 1.1) and Michael Faraday (Task 1.2) and many other well-known scientists of the 19th and early 20th centuries, had a career which embraced many different scientific topics. While working at the University of Karlruhe, he worked with BASF on the development of the Haber—Bosch Process for the synthesis of ammonia and later, while at the Kaiser Wilhelm Institute for Physical Chemistry and Electrochemistry in Berlin, he worked on combustion reactions, electrochemistry and free radical research. During the First World War, he carried out research on chlorine gas and other poison gases. He won the Nobel Prize for Chemistry in 1918 for his work on ammonia synthesis.

Carry out a web search on the history of the ammonia synthesis process and the development of the Haber—Bosch Process. Include in your search the names of Nernst, Le Chatelier and Bosch and use sources such as Wikipedia to obtain historical details of the main players in the saga. Examine also the great contribution which the synthesis of ammonia has made to the development of agriculture. (Further information on this subject can be obtained, for example, from the book by Vaclav Smil; see footnote 6.) When you have completed this task, you should have assembled a great deal more information than is contained in the main text.

Inspired by Le Chatelier's account of his experiments, Haber and Claude succeeded, less than 5 years later, in producing ammonia. Their work required the development (with the help of Robert Le Rossignol) of high-pressure equipment and also the search for effective catalysts.[6] They found that they could obtain significant yields of ammonia using catalysts based on finely

[5] http://www.platinummetalsreview.com/pdf/pmr-v27-i1-031-039.pdf

[6] A detailed account of the development of the Haber—Bosch Process and the activities of the scientists and technologists involved is to be found in the book *Enriching the Earth: Fritz Haber, Carl Bosch, and the Transformation of World Food Production* by Vaclav Smil (MIT Press, 2004) ISBN 0262693135, 9780262693134.

divided osmium, a very rare metal used at that time in light filaments.[7] Shortly after Haber had obtained his first patent in March 1909 describing the results with Os, he obtained a further patent using uranium as a catalyst; although uranium was also rare, more of this metal was available. In July 1909, Mittasch and Kranz of BASF witnessed the operation of Haber's laboratory system with Os as catalyst. In consequence, BASF initiated a crash programme to develop the process. Bosch (who had been present in Haber's lab prior to the successful demonstration but had to leave before the critical experiment was carried out) developed the high-pressure equipment required for the reactor system while Alvin Mittasch, together with his colleagues Wolf and Stern, carried out a search for a cheaper and more effective catalyst. The search was concentrated on the more abundant and cheaper elements of the periodic table, particularly on iron. Initially, none of the samples gave yields as high as those obtained with Os or U. However, in November 1909, Wolf found that one of his iron samples, magnetite (Fe_3O_4), from a mine in Gällivare, Northern Sweden, gave very high yields. Mittasch soon realized that the much improved properties were due to the presence of impurities; the first patent claiming the beneficial effects of added NaOH and KOH was issued in January 1910. Soon it was also found that MgO and Al_2O_3 were valuable promoters. The final formulation achieved, and generally still used, was based on magnetite containing 2.5–4% Al_2O_3, 0.5–1.2% K_2O, 2.0–3.5% CaO and 0.0–1.0% MgO (together with 0.2–0.5% Si present as an impurity in the metal).

BOX 1.4 Le Chatelier's Principle

At the beginning of the 20th century, chemists became engrossed by the principle of reversibility. The concept of reversibility was enshrined in Le Chatelier's Principle: that "a reaction equilibrium will move to oppose the constraints placed upon it"; in other words, the position of a chemical equilibrium can be affected by the temperature or pressure of the system. http://www.woodrow.org/teachers/ci/1992/LeChatelier.html] This applied particularly to the decomposition and formation of ammonia:

$$N_2 + 3H_2 \leftrightharpoons 2NH_3;\ \Delta H^\circ = -92.3\ \text{kJ}$$

a reaction which will proceed in either direction, depending on the conditions applied. At low temperatures and pressures, it will proceed from left to right; at high temperatures, as discussed above, ammonia decomposition into the elements will occur.

The first experimental plant for the industrial synthesis of ammonia using the Haber–Bosch Process (Bosch being included in the name as a result of his crucial contribution in the scale-up

[7] The total world supply of osmium at the beginning of the twentieth century was 100 kg and BASF acquired all of that for the sum of 400,000 marks; even then, BASF calculated that that was only enough to produce 750 tons of ammonia per year.

of the process and the introduction of the associated ground-breaking high-pressure technologies) was built in Appau and began operation in August 1910, producing 18 kg of ammonia per day by January 1911. The first commercial plant in Appau began operating in September 1913, giving 10 tons of ammonia per day. Today's world-wide production is more than 160 million metric tons per year, there being more than 600 large plants producing 1000 tons or more per day.[8] Without the supplies of ammonia and hence of the fertilizers produced from the ammonia, the world's agriculture would be incapable of sustaining the production of foodstuffs necessary for its current population. Smil[6] has estimated that only about half of today's agricultural production would be possible without the Haber–Bosch Process. Haber was awarded the Nobel Prize for chemistry in 1918 and Bosch was awarded the Prize in 1931 (sharing it with Bergius who had been responsible for the development of the process for the hydrogenation of coal).

BOX 1.5 Testing of Ammonia Synthesis Catalysts

In order to commercialise the Haber–Bosch Process, BASF urgently needed a cheaper and more abundant catalyst than the Os catalyst developed by Haber. Mittasch initiated a crash programme of the type never before attempted to find a suitable material using a series of 30 parallel high pressure mini-reactors developed by his colleague Stern. Each of these had an easily replaced cylinder containing 2 g of catalyst. This set of systems enabled the Mittasch team to test many thousand samples of catalyst over a period of two years and the work led to the development of the promoted iron catalysts of the type that are still used today. It is interesting to note that the element Ru had not been discovered at that time and it was only in 1970's that it was tested and found to give an activity 20–50 times that of the iron catalysts (A. Ozaki and K. Aika, J. Catal., 16 (1970) 97). According to Bartholomew and Farrauto, these Ru catalysts have been commercially used, beginning in the late 1990's.

The development of the Haber–Bosch Process required a combination of a series of parallel achievements: a fuller understanding of catalysis; a recognition of the importance of thermodynamic equilibrium and the principle of reversibility; and the development of high pressure and temperature engineering materials and methodologies. The award to Bosch of his own Nobel Prize recognized his major contribution in the engineering of the process without which it could not have succeeded. BASF continued to capitalize on the expertise gained and, in 1923, Bosch, Mittasch and their colleagues commercialized the process for the production of methanol from water gas derived from coal using a ZnO/Cr_2O_3 catalyst at 300–400 °C and 300 atm. pressure.[9] Also using the high pressure technology developed by Bosch and his colleagues, Fischer and Tropsch, working at the Keiser Wilhelm Institute for Coal Research at Müelheim an der Ruhr, showed in 1922 that alkalized iron filings could convert water gas, at

[8] We will return in Chapter 7 to a description of the plants used in ammonia synthesis.

[9] Only in 1966 was the alternative low-pressure process (220–300 °C and 50–100 atm. pressure) introduced by ICI Fertilizer Division in Billingham, now part of Johnson Matthey.

a temperature of 400 °C and at a pressure of more than 100 atm. pressure, to a high molecular weight oxidized hydrocarbon mixture which they called "Synthol". Later developments led to the use of iron and cobalt catalysts for the production of liquid hydrocarbons from coal-derived syn-gas, the so-called Fischer–Tropsch Process, commercialized during the war years by Ruhr–Chemie in Holten. After the war, the main use of the process was in South Africa where Sasol produced synthetic fuels from coal using the technology. Relatively recently, there has been a significant revival of interest in the process, largely for use as a step in the conversion of natural gas to fuels (GTL, gas to liquids), for example, using the Shell Middle Distillates (SMDS) Process, for use in regions of the world having large reserves of natural gas.[10] The process is also now being used in China, a country which has large reserves of coal but few natural gas or oil resources.[11]

1.2.6 Steam Reforming of Hydrocarbons

Another critical development associated with the work on all these processes was the introduction of the steam reforming of hydrocarbons, again pioneered by Mittasch and Schiller in BASF in the 1920s and 1930s. As Germany had no natural gas available, the methane needed for the experiments was synthesized from CO and hydrogen; Schiller succeeded in reforming this methane using a nickel catalyst. The first publication in the open literature of work on the conversion of methane and steam was a paper by Neumann and Jacob[12] that appeared in 1924; this showed that the gas products were close to equilibrium with respect to the steam reforming and water–gas shift reactions. A licensing agreement transferred the BASF technology to Standard Oil of New Jersey who began to produce hydrogen for use in its Baton Rouge refinery in 1931.[13] A steam reformer was commissioned by ICI in Billingham in about 1936. While natural gas continued to be the feedstock of choice in the US, light distillate naphthas became

---

[10] Bartholomew and Farrauto have described the development of FTS technologies as having five phases: (1) discovery of Co and Fe catalysts (1902–1928); (2) commercial development of the Fischer cobalt-based process (1928–1945); (3) age of iron and Sasol (1946–1974); (4) rediscovery of FTS and cobalt (1975–1990); and (5) birth/growth of the GTL industry (1990–present). (C.H. Bartholomew and R.J. Farrauto, "Fundamentals of Industrial Catalytic Processes, 2nd Edition, Wiley Interscience, 2006).

[11] Many of the developments described above took place in Germany before and during the two world wars; much of this was described in patents but there was also much secret know-how. Some of the earlier technology, such as the steam reforming process, was licensed internationally between the wars. Towards the end of the Second World War, the allies recognized that there existed in Germany a lot of technology which was not generally available and so teams of technical experts were sent to Germany (and other countries) to obtain details of processes such as the Fischer–Tropsch Synthesis Reaction. A web site exists which gives access to a great deal of this material and which also gives fascinating glimpses of the personalities involved: http://www.fischer-tropsch.org.

[12] B. Neumann, K. Jacob, Z. Electrochem. 30 (1924) 557 (referred to by J.R. Rostrup Nielsen, in: J.A. Anderson, M. Boudar (Eds.), Catalysis Science and Technology, vol. 5, Springer Verlag, Berlin, Heidelberg, New York, Tokyo 1984.)

[13] Two years earlier, Shell produced hydrogen from methane but by thermal cracking of the methane.

the feedstock used in Europe during the 1950s, this situation only changing when natural gas supplies were located in the North Sea and elsewhere in the 1960s. The development of the steam reforming of naphthas coincided with the development of new steels which could be used at high temperatures and pressures; new catalysts were also developed which had the capability of withstanding these high temperatures and of resisting carbon lay-down. Two tubular reformers capable of operating at around 15 atm. pressure were commissioned by ICI in 1962 while one operating at 40 atm. pressure was commissioned by Topsøe less than 5 years later. Steam reforming is probably still one of the most critical processes taking place in the chemical industry and improvements in the technology are continuously being reported, the aim being to increase the efficiency and hence to lower the cost of hydrogen production. We will return to the topic of steam reforming of methane and related reactions in Chapter 8, Section 2.

1.2.7 Basic Research on Catalysis

Much of the industrial development work of the early twentieth century discussed above was followed by more basic studies in universities and research institutes aimed at gaining a greater understanding of the science behind these catalytic processes. A number of names are associated with these developments: Langmuir for his work on adsorption, Taylor for his development of the idea of the active site, Emmett for his work on physical adsorption, Rideal, Tempkin, Boudart and many others for their work on the kinetics of catalytic reactions. These names will recur frequently, with others, throughout the remainder of this book, and the interested reader should carry out searches on the internet to find out more about the people behind the names. As an example, the contributions of Langmuir, a Nobel awardee in Chemistry in 1932 (*Nobel Lectures, Chemistry 1922–1941*, Elsevier Publishing Company, Amsterdam, 1966), will now be discussed briefly.

Irving Langmuir graduated in metallurgy from Columbia University in 1903 and proceeded to obtain an MA and PhD from Göttingen in 1906, working there under Nernst where he did work on the "Nernst Glower", the electric light bulb invented by Nernst. He returned to the US and took up an academic position in New Jersey which he held until 1909 when he joined the laboratories of the General Electric Company at Schenectady, New York, ending up as Associate Director.[14] His work there stemmed from his involvement in vacuum physics related to incandescent lamps and vacuum discharges (developing for that work the modern diffusion pump) and he went on, *inter alia*, to work on the behaviour of gases at low pressure.[15] He showed that hydrogen was dissociated into atoms on the incandescent filament of a light bulb

[14] Langmuir has been described as being the father of basic research in industry.

[15] Langmuir's name is associated with many other developments such as the Langmuir–Blodgett trough used to study surface films on liquids, the Langmuir probe used to measure charged particles in a discharge tubes, the so-called Langmuir waves in plasmas and the phenomenon of seeding of clouds to bring about precipitation. He was also responsible for the development of the gas-filled incandescent lamp and the discovery of atomic hydrogen.

and that it was adsorbed on the surface of the bulb as a monomolecular layer.[16] This led to the so-called Langmuir Isotherm (See Chapter 2, Section 4) to describe chemisorption on surfaces and also ultimately to the kinetic description of bimolecular catalytic process, the so-called Langmuir—Hinshelwood kinetics. The work of Langmuir and others in that period led to an understanding of the importance of chemisorbed species in catalytic processes and the concepts of surface intermediates and active sites followed soon after.

Langmuir's work on ultra-high vacuum (UHV) technology led in the 1960s to the development of important surface techniques such as field emission spectroscopy, field ion spectroscopy, low energy electron diffraction, Auger spectroscopy and X-ray photoelectron spectroscopy[17].

TASK 1.4 Ultra-High Vacuum Methods

The development of ultra-high vacuum techniques led to the development of many modern techniques for the study of the adsorption and reactions of molecules on surfaces. We will return in later chapters to more detailed discussion on a number of these methods. A listing of the modern methods available is given at: http://en.wikipedia.org/wiki/List_of_surface_analysis_methods/. From the perspective of the historical development of the subject, examine the development of the techniques and approaches used in work on adsorption, paying particular attention to techniques such as field emission microscopy (FEM), field ion microscopy (FIM), X-ray photo-electron spectroscopy (XPS, originally known as electron spectroscopy for chemical analysis, ESCA) and the work of scientists such as Erwin W. Müller, Kai M.B. Siegbahn, Gabor A. Somorjai, Gerhard Ertl, Sir David L. King and Sir John M. Thomas. You should restrict your searches to using a web browser such as Google (but see also Task 1.5).

The availability of commercial equipment based on these methods led to a period of intense activity on "Surface Science,"[18] the investigation of the reactions of molecules with atomically clean solid surfaces using these modern physical techniques. Surface science equipment (see, e.g., http://www.vgscienta.com/), requiring the use of reproducible ultra-high vacua in which surfaces may be maintained in a clean state over significant periods of time, is very expensive to purchase and also to operate. Hence, surface science has been accused of hogging the research budgets of many research councils over several decades. More recently, with the

[16] I. Langmuir, J. Am. Chem. Soc. 40 (9) (1918) 1361—1403.

[17] The 1981 Nobel Prize for Physics was awarded to the Swedish physicist Kai M. Siegbahn [http://nobelprize.org/nobel_prizes/physics/laureates/1981/] for developing this technique for studying the species adsorbed on surfaces by monitoring the secondary electrons emitted when the surface is exposed to electromagnetic radiation.

[18] As opposed to "surface chemistry", a term which is applied more to the gas—liquid and solid—liquid interfaces, although it is sometimes also applied to the solid—gas interface.

development of methods of using these techniques for studying reactions at surfaces at realistic pressures, more emphasis has been placed on the investigations of catalytic reactions. Scientists whose work is notable in this field are Gabor Somorjai and Gerhard Ertl. The former has been described as the "Father of Modern Surface Chemistry" [http://en.wikipedia.org/wiki/Gábor_A._Somorjai/] while the latter was awarded the Nobel Prize in Chemistry in 2007, in particular for having applied his approach to a fundamental examination of the mechanism of the Haber synthesis of ammonia over iron surfaces [http://nobelprize.org/nobel_prizes/chemistry/laureates/2007/ertl_lecture.pdf].

BOX 1.6 Web-based Resources

It is worth noting at this stage that work carried out until about the middle of the 20th century is most easily followed up by carrying out general web searches and that these often lead to articles in Wikipedia and other similar items. However, it is possible for searches on more recent topics to use more specific search tools such as Chemical Abstracts, Scifinder, Google Scholar, Web of Science and Scopus. In subsequent chapters, you will be asked to use citation indexing in the form of either Web of Science or Scopus very extensively in conjunction with the material of those chapters as a means of identifying the most relevant papers on a particular topic and hence of learning more on each topic. Citation searching has the great advantage over the more conventional use of abstracting resources that it allows one to search forward in time from a base reference in order to identify current work on a specific topic rather than just searching backwards.

It is worth noting that in addition to being used for "forward" literature searching, citation indexing, over the last decade or so, has been used to assess scientific output. The number of citations to a particular scientist's publications can be used as a measure of that person's scientific output. There are many limitations to such an approach, the most important being that citation behaviour varies extensively from field to field; another important factor is that reviews and "methods papers" often attract much higher numbers of citation than do more traditional scientific communications. In order to help assess output, the so-called Hirsch Index, *h*, has been introduced (J.E. Hirsch, An index to quantify an individuals's scientific scientific research output. Proc. Nat. Acad. Sci. USA. 102(2005) 16569–16572; cited > 800 times). The value of *h* is the number of papers published by a particular author which have been cited more than that number of times; in other words, if an author has published 200 papers and 30 of these have been cited 30 or more times, the appropriate *h*-index is 30. This method has the advantage that it gives less weight to reviews and methods papers and pays more attention to general output. When searching for the most significant authors on a particular topic, it can be helpful to identify the *h*-index of those in the list and then concentrate on their papers.

In order to learn more about the use of the *h*-index, it is interesting to follow up some of the papers citing the original paper by Hirsch cited above.

TASK 1.5

In order to become familiar with the use of citation methods, you should now carry out searches on some of the authors shown in Task 1.4 and identify their most important papers on some of the surface science techniques. Follow this by identifying the most important recent articles citing each of these papers; hence, identify the most significant scientists, in addition to those listed, working in the appropriate fields.

This chapter has been an attempt to introduce some of the ideas behind the subject of catalysis and to trace some of the important milestones in the development of the subject. A number of the topics introduced here will be considered in more detail later in the book, others not. The student is encouraged to follow up as many of the topics introduced as possible, for example, by using the electronic linkages provided in the text and exploring some of the wealth of information available on the web.

CHAPTER 2

Surfaces and Adsorption

Chapter Outline

2.1 Introduction

It is generally considered that there are three states of matter, solid, liquid and gas, although each of these can be further refined or subdivided. As far as you, as a chemist, are concerned, many of the reactions with which you have had experience occur in either gas or liquid phase even though there are also many reactions that occur in the solid phase. However, further thought leads one to recognize that one is very familiar with occurrences at the interface between two or even more phases: the formation of gas bubbles in a liquid, the action of a surfactant at a liquid–solid interface, corrosion at a solid–gas interface (or a solid–liquid–gas interface), friction at a solid–solid interface. All surfaces are "unsaturated" in that they can formally be thought of as being created by slicing through a bulk phase (see Figure 2.1). These unsaturated surfaces, therefore, possess surface energy (or, as it is occasionally rather loosely put, "dangling valencies"). The surface energy can contribute to the all-over energy changes occurring when a process occurs at the surfaces, whether this process is purely a physical interaction or a chemical reaction. The study of the chemical phenomena occurring at such interfaces can be loosely termed "surface chemistry" and, as noted in the previous chapter, the term can also embrace the study of the adsorption and reactions of

Heterogeneous Catalysis. DOI: 10.1016/B978-0-444-53363-0.10002-7

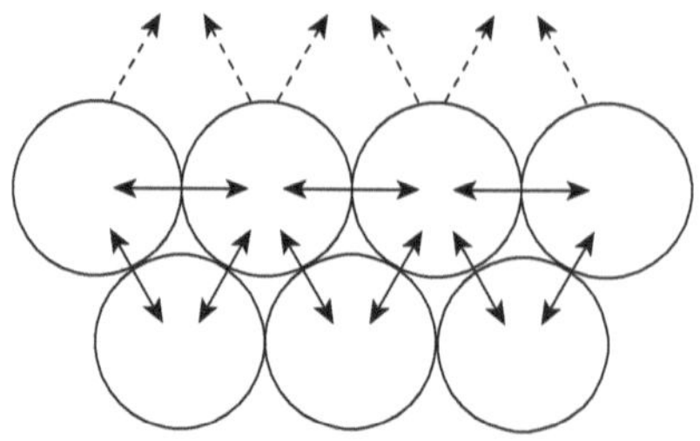

Figure 2.1:
Schematic representation of the cross-section of a surface generated by cleaving a metal crystal.

molecules at solid surfaces (normally adsorbed from the gas phase but sometimes from the liquid phase). When a reaction at a surface leads to a new surface entity that desorbs to give products and results in the regeneration of the surface, we have a catalytic process and are dealing with the phenomenon of "heterogeneous catalysis" (see Section 1.2). In order to understand heterogeneous catalysis, we should have a good understanding of the phenomena of adsorption and, to a lesser extent, desorption. This chapter aims to give the student an understanding of the factors determining the behaviour of surfaces during the adsorption (and desorption) processes and of how one would study these processes.

2.2 Clean Surfaces

In order to study the phenomenon of adsorption experimentally, it is essential, first, to be in a position to generate clean surfaces. The earliest work on chemisorption coincided with the development of the vacuum techniques necessary to generate and maintain clean surfaces (see Chapter 1). The work of Langmuir was driven by the development of the electric light, initiated during his stay in Nernst's laboratory in Göttingen. When, in 1909, he joined the laboratories of General Electric in Murray Hill, New Jersey, he started to work on the use of incandescent filaments for what we now know as the electric light bulb (Figure 2.2).

Typically, these filaments are made of thin tungsten wire which, when heated to temperatures above 2000 °C, give out light in the visible wavelength. Tungsten is one of a group of refractory metals that melt at very high temperatures, the melting point of tungsten T_m being about 3410 °C (Table 2.1). The surface atoms of a metal can become mobile at a temperature, the so-called Tamman temperature, T_t, significantly below the melting point. The Tamman temperature is given by:

$$T_t \cong 0.6T_m \qquad (2.1)$$

Hence, the surface mobility of tungsten atoms is not expected to occur until ca. 2000 °C. Tungsten metal can be oxidized in air to give WO_3. At relatively low temperatures, the

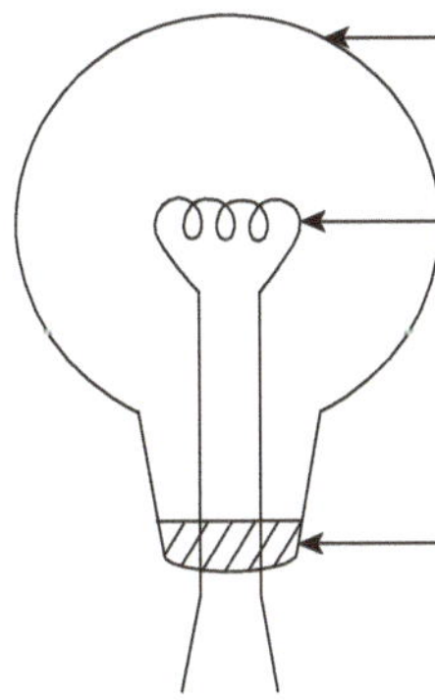

Figure 2.2:
Schematic representation of an electric light bulb.

Table 2.1: The melting points of some refractory metals.

Metal	Melting Point (°C)
Tungsten	3410 ± 20
Molybdenum	2617
Tantalum	2996
Rhenium	3180
Niobium	2468 ± 10
Osmium	3045 ± 30

Source: *Handbook of Chemistry and Physics, CRC Press.*

oxide forms a non-mobile layer that passivates the surface against further oxidation. However, as the temperature is increased further, multiple layers of oxide are first formed and then the oxide layer becomes volatile, causing desorption of WO_3.[1] As long as excess oxygen is excluded from the filament, any surface layer of oxide is desorbed relatively easily, resulting in a clean and stable tungsten surface, as long as no other gases are present. The exclusion of oxygen from the filament requires the use of high vacua and it was the development of the methods needed to create such vacua that led to the first successful incandescent light bulbs.

All early vacuum apparatus was made of glass and every laboratory using such equipment employed one or more professional glass-blowers. Further, all research students working with such systems became relatively proficient amateur glass-blowers.[2] A typical vacuum system was

1 Desorption of any surface impurities takes place at the same time; further, some surface oxygen may desorb as molecular oxygen rather than as WO_3.

2 In the late 1960s, I worked in a laboratory where we had, as our senior glass-blower, Harry Pierce who had made magnetrons, the main components of the radar equipment used during the Second World War. He could make or repair anything, often much to my relief, as my own proficiency as a glass-blower did not stretch to mending diffusion pumps.

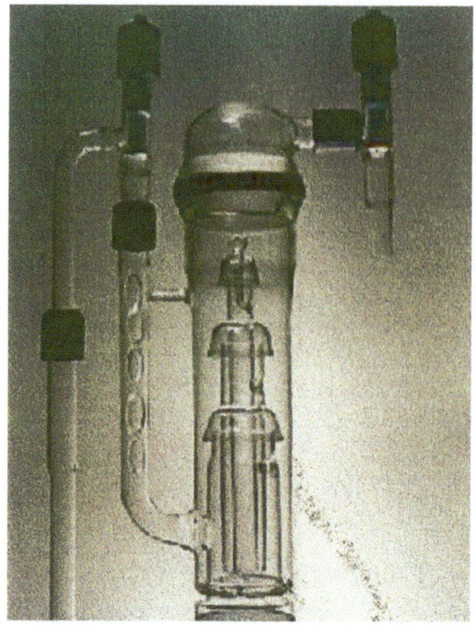

Figure 2.3:
A modern glass oil diffusion pump. Courtesy of Michael D. Wheeler of Arizona State University. http://www.public.asu.edu/~aomdw/GLASS/DIFFUSION_PUMP.html.

made up of a rotary pump, which would give a vacuum of the order of 10^{-2} mm Hg,[3] a mercury diffusion pump, which would give, with suitable techniques, a vacuum as low as 10^{-9} mm Hg,[4] various vacuum taps (greased stop-cocks, mercury cut-offs or, rather later, metal valves joined to the system by suitable metal–glass seals) and pressure measuring devices including mercury manometers, McLeod gauges and ion gauges. In order to achieve the lowest possible pressure, which was dependent on the vacuum components such as the pumps and valves used, elaborate preparation methods, such as bake-out and trapping at low temperatures to exclude mercury vapour, were necessary. More modern equipment uses oil diffusion pumps, ion pumps or turbomolecular pumps. Although the first oil diffusion pumps were made of steel, glass versions have also been made. Figure 2.3 shows a glass oil diffusion pump developed at Arizona State University and Figure 2.4 shows a glass reaction system incorporating such a pump.

TASK 2.1 Vacuum Techniques

Do a web search on vacuum techniques, diffusion pumps, pressure gauges, etc., in order to get an idea of the techniques involved and the suppliers of such equipment.

---

[3] The unit mm Hg was a very practical one, being measured directly using a mercury manometer or, less directly, using a McLeod gauge. The term Torr (from Torrocelli, the Italian scientist who discovered the barometer in 1644) is also often used. 1 mm Hg or Torr is equivalent to 13,341 Pa. (For more details on pressure units, see http://en.wikipedia.org/wiki/Torr/)

[4] Langmuir was responsible for the introduction of a modified mercury diffusion pump, which still bears his name.

Figure 2.4:
A typical modern glass vacuum apparatus made and used at Arizona State University (Courtesy of Michael D. Wheeler; see Figure 2.3).

2.3 Langmuir's Work on Adsorption

During his work on the light bulb, Langmuir discovered that hydrogen could interact at a hot tungsten filament to give hydrogen atoms. He also recognized that hydrogen could become *ad*sorbed on the tungsten surface, this process proceeding through the dissociation of the hydrogen–hydrogen bonds. The word *adsorbed* implies "uptake on" the surface of a solid; when the interaction is chemical, it is referred to as "chemisorption", but when the interaction is purely physical, it is referred to as "physical adsorption" (or "physisorption"). As both chemisorption and physical adsorption play very important roles in relation to the phenomenon of heterogeneous catalysis, we will now consider each in detail.

The phenomenon of chemisorption has received most attention in relation to metal surfaces. Langmuir's early work was carried out using clean metal filaments produced under high vacua. This led to further experimentation using evaporated metal films (generally prepared under essentially what we now recognize as "ultra-high vacuum" (UHV) using laboratory-built glass equipment) and, later, to work using single crystals in purpose-built commercial stainless steel UHV systems.

2.4 The Langmuir Isotherm[5]

Langmuir's pioneering work allowed him to develop the well-known Langmuir adsorption isotherm. This equation can be used to describe the adsorption of a gas on a clean surface, generally treated as being a metal (although the arguments could be modified to apply equally to oxides and other materials).

[5] In this chapter, references are not provided to the original literature in relation to material that has become "bookwork." In other words, the material is so well established that the original references are seldom read. However, the student can perform searches to access the original papers, if so desired, remembering that some of them may be in relatively inaccessible journals or in languages other than English.

The simplest form of the Langmuir adsorption isotherm describes the equilibrium adsorption of atoms A from the gas phase on the sites (M) of a metal surface at a constant temperature:

$$M + A_{(g)} \leftrightarrows A - M \tag{2.2}$$

The resultant equation (see Box 2.1 for derivation) is:

$$V_{ads}/V_m = \theta_A = b_A P_A/(1 + b_A P_A) \tag{2.3}$$

BOX 2.1 Derivation of the Langmuir Adsorption Isotherm

Consider a surface made up of n_S equivalent surface sites (e.g. metal atoms). If a gas A is in equilibrium contact with that surface and some A is adsorbed on the surface so that the surface concentration is n_A, there will be $(n_S - n_A)$ empty sites.

Assuming that adsorption is not activated, the rate of adsorption of A on the surface, r_{ads}, will be given by:

$$r_{ads} = k_{ads} P_A (n_s - n_A) \tag{2.4}$$

and the rate of desorption from the filled sites, r_{des}, will be given by:

$$r_{des} = k_{des} n_A \tag{2.5}$$

At equilibrium, $r_{ads} = r_{des}$; and so

$$r_{ads} = k_{ads} P_A (n_S - n_A) = r_{des} = k_{des} n_A \tag{2.6}$$

Dividing through by n_S and replacing n_A/n_S by θ_A, we obtain:

$$k_{ads} P_A (1 - \theta_A) = k_{des} \theta_A \tag{2.7}$$

or

$$\theta_A/(1 - \theta_A) = k_{ads}/k_{des} P_A = b_A P_A \tag{2.8}$$

where b_A is the so-called adsorption constant. The value of b_A depends on the activation energies for adsorption and desorption; in other words, it is directly related to the heat of adsorption.

Rearranging gives:

$$\theta_A = b_A P_A/(1 + b_A P_A) \tag{2.9}$$

This is the Langmuir isotherm describing the adsorption of an atom A on a uniform surface.

It applies to the adsorption of an undissociated species A on the atoms of a metal surface at a constant temperature, these metal "surface sites" being either empty (M) or occupied by the species A "adsorbed" on the site ($A - M$). Here, V_{ads} is the volume of A adsorbed, V_m is the volume of A required to give a full monolayer, θ_A is the fractional coverage of the surface by A, b_A is the so-called adsorption constant and P_A is the pressure of A. (It should be noted that the amounts adsorbed can be expressed in many ways. The volume of gas adsorbed (measured at

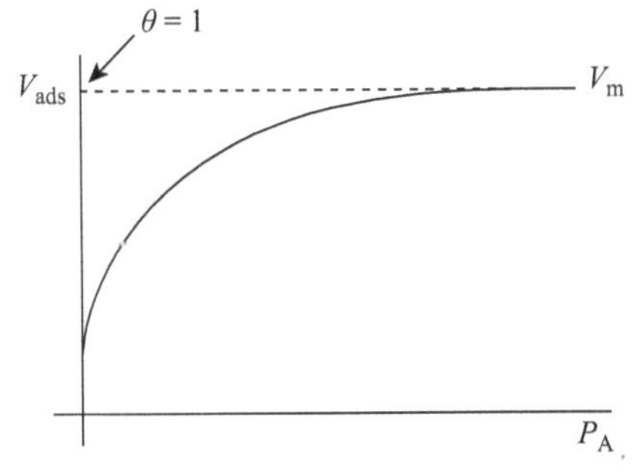

Figure 2.5:
Typical adsorption isotherm following the Langmuir equation; when the volume adsorbed (V_{ads}) is equal to the monolayer capacity (V_m), no further adsorption occurs.

STP) is frequently used; however, this quantity could equally well be converted to the number of molecules adsorbed, n_A, or the mass of gas adsorbed, m_A. The derivation of the equation shown in Box 2.1 depends on all the sites being identical; in other words, the heat of adsorption of the hydrogen is unchanged as the surface fills up. As b_A is related to the heat of adsorption ($b_A = a.\exp(\Delta H_{ads}/RT)$), this implies that b_A is independent of the coverage of the surface by A. As we shall see, this is far from the real situation. Nevertheless, the isotherm is of great importance as it allows us to gain a better picture of the adsorption/desorption equilibrium that exists and it allows us to proceed to apply the same approach to actual catalytic processes (see Chapter 6).

Figure 2.5 shows the typical form of an isotherm for the adsorption of molecule A on a surface. There is no adsorption at $P_A = 0$, while the volume adsorbed approaches the value V_m asymptotically as the pressure increases. In other words, the sites are gradually filled until there are no empty sites remaining; that is once V_{ads} reaches V_m, the so-called monolayer capacity of the surface. In handling data of this sort, it is helpful to be able to linearize the plots in order to obtain a more accurate value of V_m and to allow the constant b to be determined. One such linearization (can you find another?) is to plot the data simply as reciprocal values as shown in Figure 2.6.

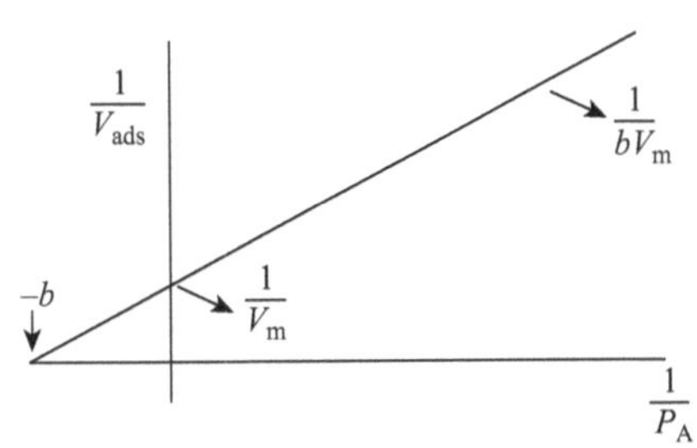

Figure 2.6:
Plot of the reciprocal form of the Langmuir equation from which V_m and b_A may be determined.

The value of V_m is obtained from the y intercept and the "adsorption constant" b is obtained from the negative intercept on the $1/P_A$ axis.

TASK 2.2 Information from the Reciprocal Plot of Figure 2.6

(a) The value of b can also be obtained from the gradient of the plot. How?
(b) How can one determine the heat of adsorption of A on the surface of the adsorbent for the data of Figure 2.5. (Hint: you would need to obtain further data).
(c) What would be the form of the plots in Figures 2.5 and 2.6 if the heat of adsorption was much lower.

TASK 2.3 Determination of the Metallic Surface Area of a Sample

If the value of Vm determined for the non-dissociative adsorption of a species A on a 0.15 g of a metal is 1.7 cm^3 (STP)*, determine the specific surface area of the metal assuming that one A occupies 0.65 nm^2 on the surface.

* STP = standard temperature and pressure; one mole of any ideal gas occupies 22.4 dm^3 at STP.

As was shown in Box 2.1, the constant b_A is given by k_{ads}/k_{des} and is hence related to the heat of adsorption ΔH_{ads}: the higher the value of the heat of adsorption (or the greater the strength of the M–A bond formed), the higher is the value of b_A. The shape of the isotherm is strongly dependent of the value of b_A. For low heat of adsorption, the isotherms have poorly defined "knees" and the asymptotic value of V_m is difficult to determine; in contrast, for high values of the heat of adsorption, the "knee" is well defined and it is possible to determine the value of V_m purely by inspection. The shape of the isotherm is also determined by the adsorption temperature: an isotherm with a diffuse knee obtained for adsorption at a relatively high temperature can become relatively sharp at lower temperatures. We will return to the topic of the temperature of adsorption in Section 2.8 in relation to the subject of the adsorption of more complex molecules.

The isotherm shown in Figure 2.5 has two important features: at low pressures, the volume adsorbed is proportional to P_A while at high pressures, the volume adsorbed is independent of pressure ($\theta \approx 1.0$). (This has consequences in relation to the rates of catalysed reactions (see Chapter 6); a unimolecular catalysed reaction can have kinetics ranging from first order to zero order, depending on the coverage of the catalyst surface by the reactant.) In practice, it is also often found for intermediate coverages that

$$\theta_A \approx b_{A^n} P_{A^n} \tag{2.10}$$

where $0 < n < 1$. This empirical equation is known as the Freundlich Isotherm (see further below).

2.5 The Chemisorption of Hydrogen

For the case of the adsorption of hydrogen, the form of the Langmuir isotherm has to be modified to allow for dissociation of the hydrogen. The equation (see Box 2.2 for derivation) now contains terms containing the square root of the hydrogen pressure and the constant c has a different relationship to the heat of adsorption:

$$\theta_{\mathrm{H}} = \frac{C_{\mathrm{H}}\, P_{\mathrm{H}_2}^{0.5}}{1 + C_{\mathrm{H}}\, P_{\mathrm{H}_2}^{0.5}} \tag{2.11}$$

For intermediate coverages,

$$\theta_{\mathrm{H}} = CH^{n} P_{\mathrm{H}_2^{n}} \tag{2.12}$$

where again $0 < n < 1$.

BOX 2.2 Dissociative Hydrogen Adsorption – The Pressure Function

When a molecule such as hydrogen dissociates on adsorption, we can no longer use the simple Langmuir Isotherm to describe the adsorption behaviour. As we shall see later (Section 2.9), dissociative adsorption takes place following physical adsorption of the H_2 molecule. It is possible to set up the corresponding Langmuir isotherm using a similar approach to that shown in BOX 2.1. An alternative method is to consider the existence of a *virtual equilibrium* involving the formation of hydrogen atoms in equilibrium with molecular hydrogen in the gas phase as shown in the following scheme:

$$\begin{array}{ccc} \mathrm{H2\ (g)} & \leftrightarrows & \mathrm{2\ H(g)} \\ \downarrow\uparrow & & \downarrow\uparrow \\ \mathrm{H2_{(ads)}} & \leftrightarrows & \mathrm{2\ H_{(ads)}} \end{array}$$

As long as the system is in equilibrium, the *virtual pressure* of hydrogen atoms so formed determines the adsorption behaviour.

For the gas phase (virtual) equilibrium, we can write:

$$\mathrm{K_{eq}} = \mathrm{P_{H^2}/P_{H2}}$$

from which we obtain:

$$\mathrm{P_H} = (\mathrm{K_{eq}.P_{H2}})^{0.5}$$

Inserting this expression for P_H instead of P_A in Eq. 2.2 (with $c_H = b_H.K_{eq}{}^{0.5}$) gives us Eq. 2.4. In practice, it is not of importance that the gas phase equilibrium is unlikely to be achieved as long as the alternative route via adsorbed molecular hydrogen is fully equilibrated.

TASK 2.4 Adsorption Constants for Complex Molecules

The adsorption constants for more complex molecules are no longer simply related to the heat of adsorption as shown above for the adsorption of A. See if you can work out the relationships, which now apply for each of the cases given above.

2.6 The Chemisorption of More Complex Molecules

The approach outlined above in Box 2.2 can also be applied to more complex situations. For example, if one considers that the adsorption of methane occurs on a metal surface according to the following equilibrium reaction:

$$M + CH_{4(g)} \leftrightarrows CH_3 - M + H - M \tag{2.13}$$

it can be shown that:

$$\theta_{CH_3} = kP_{CH_4}/P_{H_2^{0.5}}/(1 + kP_{CH_4}/P_{H_2^{0.5}}) \tag{2.14}$$

If we now use the approximation for intermediate coverages,

$$\theta_{CH_3} = k'(P_{CH_4}/P_{H_2^{0.5}})^n \tag{2.15}$$

where k' is a new constant and $0 < n < 1$.

TASK 2.5 Langmuir Isotherm for the Adsorption of Methane

Example

Derive the isotherm and the resultant approximation for the situation in which adsorption of methane occurs by the equilibrium:

$$M + CH_{4(g)} \leftrightarrows CH_2{=}M + 2H - M \tag{2.16}$$

Answer

$$\theta_{CH_3} = kP_{CH_4}/P_{H_2}^{1.0}/(1 + kP_{CH_4}/P_{H_2}^{1.0}) \tag{2.17}$$

$$\theta_{CH_3} = k'(P_{CH_4}/P_{H_2})^n \tag{2.18}$$

This approach can be applied to even more complex situations. For example, see if you can derive an isotherm for the adsorption of ammonia to give nitrogen and hydrogen "ad-atoms".

2.7 Non-homogeneous Surfaces

It is well documented that the heat of adsorption of a molecule on a surface seldom remains constant as a function of coverage; instead, the heat of adsorption almost invariably decreases as coverage increases. Various attempts have been made to describe the adsorption behaviour of such cases. Examples of such approaches are the Freundlich and Temkin equations. Although the former equation (Eq. 2.9) was initially seen as being entirely empirical, Zeldovich showed that it could be derived for a situation in which the heat of adsorption falls logarithmically with coverage, the resulting linear form being:

$$\text{In}\ \theta = RT/Q_0.\text{lnp} + \text{constant} \tag{2.19}$$

where Q_0 is the heat of adsorption at zero coverage. This reduces to the more familiar form of the Freundlich equation (see above):

$$\theta = kp^{1/\text{m}}$$

where m is greater than unity, this being equivalent to Eq. 2.10.

The so-called Temkin equation

$$\theta = \text{a}.ln(\text{b}p) \tag{2.20}$$

where a and b are constants whose values depend on the enthalpy of adsorption at zero coverage, was first derived by Slygin and Frumkin, but was popularized by Temkin and Pyzhev in connection with their work on the decomposition of ammonia over platinum and tungsten surfaces. It is based on the assumption that the heat of adsorption drops linearly with increasing coverage. This assumption is close to the situation which is met practically in many cases.

Although these two isotherms describe the adsorption behaviour of many gases better than does the Langmuir isotherm, the latter is more frequently used in connection with the derivation of rate equations (see Chapter 6).

2.8 Non-equilibrium Adsorption

In all that we have discussed until now, the assumption has been made that we are dealing with an equilibrium situation. After all, the term "isotherm" implies an equilibrium at a constant temperature. However, not all adsorption processes occur at equilibrium and it can be that adsorption gives rise to a species that cannot be removed reversibly. In such a case, the chemisorption process (or its reverse, i.e. desorption) is "activated"; in other words, there is an activation energy associated with one or more of the steps leading to chemisorption or desorption and so these processes may not reach equilibrium. In some cases, the adsorbed species can desorb once again under suitable circumstances. However, the adsorbed species often transforms to some other species prior to, or during, desorption; this sometimes requires elevated temperatures. An example of such complex behaviour is the adsorption of methane on various metals and this will now be discussed.

Experiments carried out on the adsorption of methane on evaporated films of tungsten prepared under UHV (a new film was used for each experiment) showed that methane could adsorb irreversibly at temperatures as low as 173 K. (The amount adsorbed corresponded to approximately a single monolayer.) When the surface covered by the adsorbed methane was heated, no gaseous product was found until a temperature of ca. 300 K, at which temperature hydrogen desorbed. At a temperature of 293 K, the average composition of the surface layer was CH_2.[6] The question remains: was all the hydrogen remaining on the surface at any temperature associated with carbon (i.e. as CHx_{ads}); was it adsorbed separately on the surface ($C_{ads} + xH_{ads}$); or was there some intermediate situation involving some free hydrogen and some CH species? The question was answered (at least partially) by carrying out "exchange experiments" with the adsorbed layer: deuterium was added to the system after heating to various temperatures and the formation of HD and H_2 was monitored. Immediately after adsorption of methane at 193 K, one hydrogen atom per adsorbed methane could be rapidly exchanged with deuterium at the same temperature; when the sample was then heated, there was a gradual exchange of more hydrogen until four hydrogen atoms had been exchanged at a surface temperature of about 300 K. Assuming that the rapid exchange corresponded to free hydrogen adsorbed on the surface and that the slow exchange corresponded to exchange with a CH_x species, it could be concluded that the methane adsorbed on tungsten at 173 K according to the equation:

$$CH_4(g) + 2W \rightarrow CH_3\text{-}W + H - W \tag{2.21}$$

On heating the adsorbed species to a higher temperature, the results indicated that the CH_3-W species decomposed further and it was found that at 373 K, the adsorbed entities were more probably CH-W species. While these conclusions are only semiquantitative (since they depend on the assumptions relating to the rate of exchange of different types of species and also relate to average surface compositions), they indicate clearly that the way in which a simple molecule such as methane decomposes when adsorbed on a metal surface is highly dependent on the conditions and that kinetic processes are quite important. Further, not all metals behave in the same way. When the experiments were repeated using Pd films, it was found that methane did not adsorb until 400 K and that it did so according to the equation:

$$CH_4(g) + Pd \rightarrow CH_3 - Pd + {}^1\!/_2 H_2(g) \tag{2.22}$$

In the case of Pd, there was no evidence for any further dissociation of the adsorbed CH_3 species. We will return to this point in a later chapter when we consider the reactions of hydrocarbons over Pd.

Another example of the adsorption of a "complex" molecule will now be considered: that of H_2S on a metals such as Fe and Pd. The H_2S molecule adsorbs at temperatures as low as 193 K on both metals without desorption of any hydrogen. However, the behaviour of the two metals differs considerably if the temperature is raised to ca. 300 K. With iron,

[6] P.G. Wright, P.G. Ashmore, C. Kemball, Trans. Faraday Soc., 54 (1958) 1692; J.R.H. Ross, M.W. Roberts, C. Kemball, J. Chem. Soc., Faraday Trans. I, 68 (1972) 221.

hydrogen is desorbed, whereas, with Pd, little or no hydrogen is found. At even higher temperatures, in the presence of excess H_2S, "sulfidation" of both metals occurs so that the equivalent of multiple layers ("multilayers") of sulfide forms. (Equivalent results are found for the chemisorption of oxygen: bulk oxidation is found to occur if the temperature is above room temperature). The stoichiometry of the sulfidation reaction with iron follows the equation:

$$Fe + H_2S \rightarrow FeS + H_2 \tag{2.23}$$

However, with Pd, another complication occurs: some of the hydrogen (in the form of atoms) is "absorbed" in the interstices of the metal to form a hydride rather than desorbing to the gas phase. The degree to which this can occur depends on the thermodynamics of the Pd–H system and to whether or not the hydrogen pressure resulting from the reaction exceeds the equilibrium pressure for the so-called ($\alpha + \beta$) phase. It is, however, interesting to note that the formation of a surface layer of adsorbed sulfur species on the Pd is sufficient to prevent the dissociation of any molecular hydrogen added to the system; hydrogen does not dissociate on the sulfur-covered surface and so cannot enter the hydride phase. Hence, it can be concluded that the twin sites necessary on the metal surface to bring about the dissociation process have all been blocked by the adsorption of sulfur.[7]

BOX 2.3 Metal Hydrides

A fuller discussion of hydrides and their formation is beyond the scope of this text, but interested readers should carry out a suitable literature search. Relevant web addresses are: http://en.wikipedia.org/wiki/Palladium_hydride/ and http://www.platinummetalsreview.com/pdf/pmr-v15-i1-021-025.pdf.

2.9 The Process of Adsorption

As described earlier in Section 2.5, hydrogen adsorbs on a metal surface in the form of atoms. However, as we discussed, hydrogen atoms are not likely to be present at any appreciable concentration in the gas phase. Then, how does molecular hydrogen dissociate at a metal surface?

The process of adsorption can be described using the so-called Lennard–Jones potential energy diagram shown in Figure 2.7. This diagram shows the potential energy of hydrogen

[7] However, it was found that if the hydrogen was dissociated on an incandescent filament in the proximity of the Pd film, the hydrogen atoms formed could be absorbed in the metal lattice; the hydride equilibrium pressure was achieved when the hot filament was turned off indicating that the desorption process was not hindered by the adsorbed sulfur layer.

BOX 2.4 Theoretical Approaches to Adsorption

The approach outlined in this text is "classical", in that it only provides a phenomenological description of the adsorption process. Modern computational methods allow much more precise descriptions to be given, not only for adsorption processes, but also for relatively simple catalytic processes. With the availability of steadily increasing computer power, the complexity of the systems studied is also increasing steadily. In essence, the methods now available allow detailed prediction of the steps that will occur when a molecule approaches a surface and when it undergoes transformations or catalytic processes. However, details of the application of computational methods to catalysis are beyond the scope of this text. Interested readers should carry out citation searches on the work of M. Neurock and his group from the University of Virginia and use these as a lead into other work of the same sort.

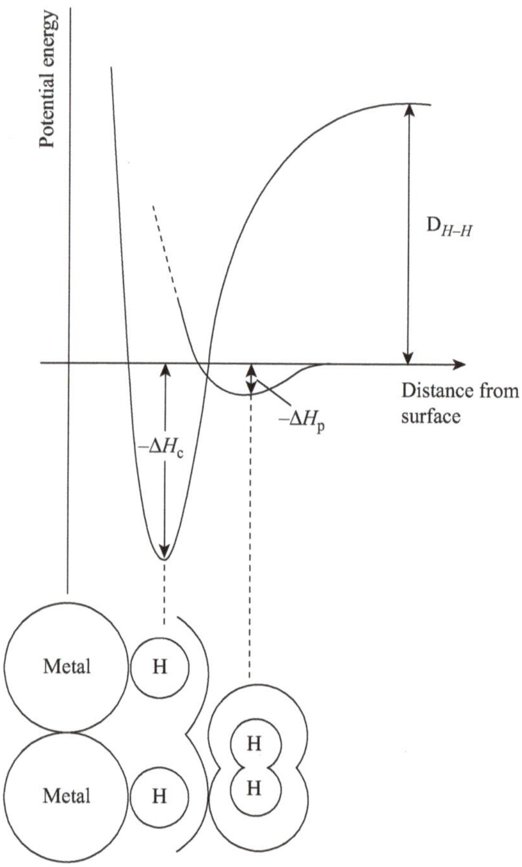

Figure 2.7:
Lennard–Jones potential energy diagram for the adsorption of hydrogen on a metal surface; as the hydrogen molecule approaches the surface from the right-hand side, it becomes physically adsorbed and then, following dissociation in the region of the surface, hydrogen atoms are chemisorbed on the surface.

molecules and hydrogen atoms as a function of the distance of each of these from the metal surface. Consider first the approach to the surface of a hydrogen molecule. At an infinite distance from the surface, there is no attraction between the molecule and the surface; the molecule is at zero potential energy. However, as the molecule begins to get closer to the surface, Van der Waals attractive forces occur and the molecule becomes "physically adsorbed" (i.e. held by physical forces of attraction, $-\Delta H_p$) at a distance from the surface corresponding to the sums of the Van der Waals radii of the hydrogen atom and the Ni surface atom, $r_{Me(VdV)} + r_{H2(VdV)}$. (If the metal is Ni, the distance is about 0.32 nm; for the other transition metals, the values will be only slightly different.) If the molecule approaches the surface any more closely, there will be a strong repulsion between the electron clouds of the molecule and the surface.

Now, consider the approach of a hydrogen atom. In order to dissociate a hydrogen molecule into two hydrogen atoms, it would be necessary to provide the dissociation energy for the hydrogen molecule, D_{H-H}, of 434 kJ. This dissociation is most unlikely to occur at any appreciable rate in the gas phase. If, however, we consider that a gaseous hydrogen atom if formed will, as it approaches the surface, feel a much greater force of attraction than the hydrogen molecule and will be able to form a metal–hydrogen bond, hence becoming adsorbed with a heat of adsorption $-\Delta H_c$ (chemisorption curve). The equilibrium position for this chemisorbed state is at a distance equal to the sum of the atomic radii of the hydrogen atom and the metal, $r_{Me} + r_H$. (For Ni, the distance is about 1.6 nm, but the values for the other transition metals will be similar). It can be seen that the chemisorption curve intersects the physical adsorption curve at a value below or very close to the zero of the energy axis. For the transition metals, the intersection generally occurs as here below the zero of potential energy and so the chemisorption process is not activated. (In other words, there is no activation energy barrier to be overcome as the hydrogen molecule approaches the surface and it transfers smoothly from the physically adsorbed state to the chemisorbed state.)

BOX 2.5 Activated Adsorption

In early textbooks describing chemisorption, it was often stated that chemisorption can be activated. However, there are few, if any, instances where that is the case for a clean surface. If activated adsorption did occur, the point of intersection of the two curves would be above the zero potential energy line. It is quite possible that the early observations of chemisorption, which led to these statements were made on "dirty" surfaces which were contaminated by partial or even complete monolayers of adsorbed oxygen. It was only with the advent of "UHV experimental adsorption systems that it became possible to maintain clean surfaces prior to adsorption measurements. We will return to the topic of UHV systems in relation to experimental methods of studying catalysts and catalytic reactions.

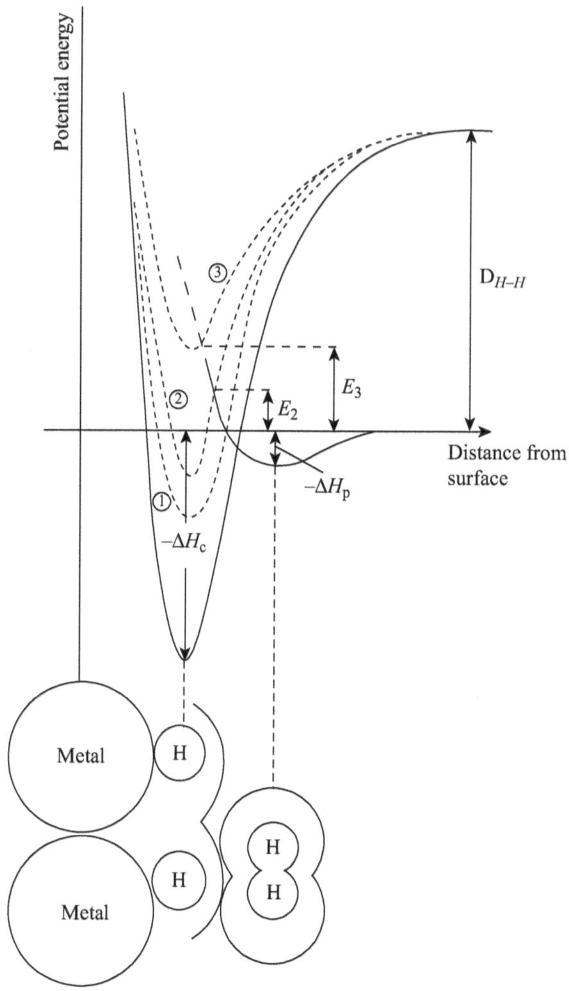

Figure 2.8:
Extension of the Lennard–Jones potential energy diagram of Figure 2.7 showing the effect of increasing the coverage of hydrogen on the surface (curves 1, 2, 3); the decrease of heat of adsorption (ΔH_c) with coverage can result in the occurrence of an activation energy for adsorption. If desorption from the adsorbed state can occur, the activation energy for desorption becomes ($\Delta H_c + E$), the value of this decreasing with increasing coverage.

The situation may however change as the surface becomes covered by hydrogen as shown in Figure 2.8: the minimum of the chemisorption curve decreases in magnitude (dotted curves 1, 2, 3), with the consequence that the curves intersect above the energy zero. This leads to the occurrence of an activation energy for the adsorption process. (We will return to this concept in a later chapter when discussing the phenomenon of catalysis.) If the adsorbed hydrogen is now

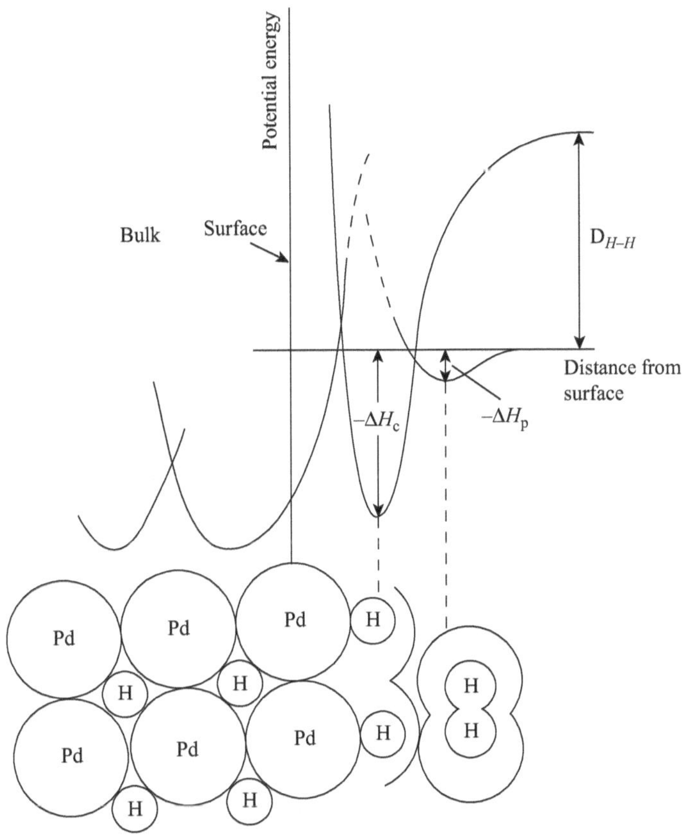

Figure 2.9:
Extension of the Lennard–Jones potential energy diagram to show the absorption of hydrogen atoms within a metal such as palladium.

to desorb to the gas phase, becoming re-associated with another hydrogen atom to form a hydrogen molecule, the reverse path has to be followed. At low coverages, the activation energy for desorption will be ΔH_c; for higher coverages, it will be ΔH_c plus the value of the activation energy for adsorption.

It is worth considering the use of such curves to describe the phenomenon of *ab*sorption discussed above in relation to the Pd/H_2 system. Hydrogen can become adsorbed on the surface of the Pd in exactly the same way as described above. However, as shown in Figure 2.9, the curve for the chemisorbed hydrogen atom now intersects with an additional curve representing hydrogen within the bulk of the metal (in interstitial sites within the metal structure); in this case, there are a series of minima extending into the bulk of the metal, each representing the most stable position for an interstitial hydrogen atom. As long as the chemisorbed curve intersects with the absorption curve at or near the energy zero, hydrogen from the gas phase can be

BOX 2.6 Precursor States

It is possible that the atom *A* can adsorb on several different sorts of sites on a single surface. Adsorption of *A* may occur on one site (Site X) and the adsorbed species may then migrate to a second more stable site (Site Y). In that case, there will be curves in the Lennard–Jones potential energy diagram corresponding to each of these. The adsorbed species at *A* can then transfer to *B*; the rate at which this occurs being related to the energy barrier to be overcome and being dependent on the temperature. If there is no energy barrier between Site X and Site Y, then all the adsorbed species will immediately go to Site Y as long as that is more energetically favoured. An alternative would be if A_2 can dissociate in two ways, to give *P*ads and *Q*ads. Then, *P*ads can be a precursor to the final stable adsorbed species *Q*ads or vice versa.

transferred seamlessly into the Pd bulk.[8] Diffusion into the bulk may itself be activated and the slow step may be either entry into the first layer of the bulk or diffusion through the Pd lattice.

2.10 Some Generalizations on Chemisorption

In the preceding sections, we have considered the adsorption and absorption of hydrogen on metals and have also discussed the adsorption of other molecules such as methane, oxygen and hydrogen sulfide. It is important to recognize that virtually every molecular species can adsorb on one or another surface and that the surfaces are not restricted to being metals: oxides, sulfides, various forms of carbon, as well as many other types of solid can all also chemisorb molecules. While there are some clear regularities in the behaviour of different molecules on different solids, each combination must be considered in its own right. The essential factor determining whether or not chemisorption occurs is that the reaction which is occurring has a negative value of ΔG. As ΔG_{ads} is given by the equation:

$$\Delta G_{ads} = \Delta H_{ads} + T\Delta S_{ads} \tag{2.24}$$

The values of ΔH_{ads} and ΔS_{ads} determine the situation. If we are dealing with the dissociative adsorption of a molecule A_2 at the surface to give $2A_{ads}$, the value of ΔS_{ads} will always be positive since the all-over system gains entropy. Hence, adsorption will only occur if ΔH_{ads} is negative. For non-dissociative adsorption (e.g. of CO on a metal), the entropy change could be negative. Hence, in principle, the adsorption of CO could involve an endothermic process. However, in practice, there is no evidence for such a situation.

[8] Originally, it was thought that the hydride of palladium could only be formed electrochemically. That was because it was necessary to achieve atomically clean Pd surfaces before the dissociation of molecular hydrogen, necessary for the subsequent absorption step, could occur; as received Pd surfaces is invariably covered by species such as adsorbed oxygen or other impurities. This is consistent with the observation mentioned above that an adsorbed layer of sulfur on the surface prevented absorption of hydrogen: the sulfur (or any other contaminant) blocks the surface sites to such an extent that the dissociation of hydrogen cannot occur.)

Relatively early in the history of the study of chemisorption, much effort was devoted to determining the heats of adsorption and attempting to find correlations between the heats measured and other properties of the adsorption system. The most successful approach was one in which the heats of adsorption were found to be relatively well correlated with the heats of formation of the equivalent bulk compounds. An example is the adsorption of oxygen on the transition metals, for which the heat of adsorption could be reasonably well correlated with the heat of formation of the most stable oxide (see Figure 2.10).

Once a reasonable corpus of data had been accumulated, a "chemistry" of chemisorption was established. For example, B.M.W. Trapnell listed the metals which would adsorb various different molecules (Table 2.2).

It is interesting to note that gold, generally considered as being very inert, is listed as chemisorbing O_2, acetylene and ethylene, but not H_2, CO_2 or N_2. The catalytic properties of gold are currently receiving much attention. We will return to that subject in a later chapter.

As mentioned in Box 2.4, computational methods are now successfully applied to investigating the way in which a molecule interacts with a surface and to calculating heats of adsorption. However, a full treatment of the subject is beyond the scope of this book.

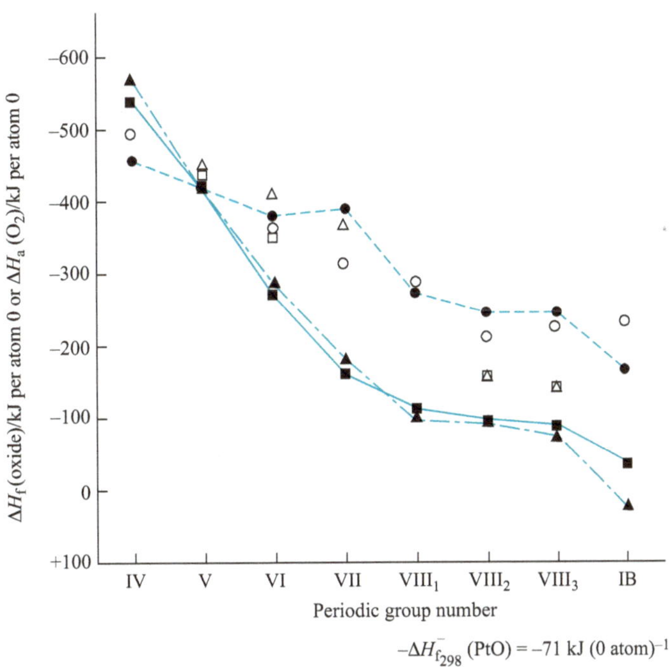

Figure 2.10:
Enthalpies of formation of oxides (ΔH_f(oxide)) (filled points) and of chemisorption on metals (ΔH_a (O_2)) as a function of the periodic group number. Circles: first transition series; squares, second transition series; triangles, third transition series.

Table 2.2: A classification of metals according to their abilities in chemisorption.

Group	Metals	Gases						
		O_2	C_2H_2	C_2H_4	CO	H_2	CO_2	N_2
A	Ti, Zr, Hf, V, Nb, Ta, Cr, Mo, W, Fe, Ru, Os	+	+	+	+	+	+	+
B_1	Ni, Co	+	+	+	+	+	+	−
B_2	Rh, Pd, Pt, Ir	+	+	+	+	+	−	−
B_3	Mn, Cu	+	+	+	+	±	−	−
C	Al, Au	+	+	+	+	−	−	−
D	Li, Na, K	+	+	−	−	−	−	−
E	Mg, Ag, Zn, Cd, In, Si, Ge, Sn, Pb, As, Sb, Bi	+	−	−	−	−	−	−

+ means strong chemisorption occurs; ± means it is week; − means unobservable.

2.11 Physical Adsorption[9]

Physical adsorption is closely related to the phenomenon of the condensation of a vapour on a surface to form a liquid layer. A first layer of molecules adsorbs on the surface and then additional layers of molecules adsorb on top of this layer in a manner similar to that observed in the condensation of a liquid. Physical adsorption is frequently studied at temperatures not far from, or equal to, the boiling point of the corresponding liquid. The most commonly encountered example encountered in work related to catalysis is the physical adsorption of molecular nitrogen on a high surface-area solid at the temperature of liquid nitrogen (78 K), its boiling point in an open vessel at atmospheric pressure. However, other molecules have also been used in specific cases, examples being the use of the adsorption of Kr or Xe at 78 K to determine low surface areas. The physical adsorption of CO_2 has also been used to determine the surface areas of zeolites.

The most widely used theory of physical adsorption was derived by Brunnauer, Emmett and Teller in 1938, the so-called BET theory. Emmett and his research group had for some time been studying iron catalysts for ammonia synthesis and had started to develop physical adsorption methods to distinguish between one catalyst formulation and another. The adsorption of gases such as nitrogen on surfaces at low temperature had been recognized for some time. However, because the isotherms obtained for different solids had very different shapes (see Figure 2.11), it was difficult to compare one isotherm with another. The BET

[9] Physical adsorption is often discussed in detail in textbooks on heterogeneous catalysis before any mention is made of the phenomenon of chemisorption. Apart from mentioning that physical adsorption is a precursor to chemisorption, I have chosen to introduce it after chemisorption for the reason that a detailed understanding of physical adsorption is less essential in understanding heterogeneous catalysis. Nevertheless, physical adsorption is important as it allows one to check the total area of a catalyst and also to find out about its pore structure. Your instructor may prefer to handle this section in connection with the material of Chapter 4 on catalyst characterization.

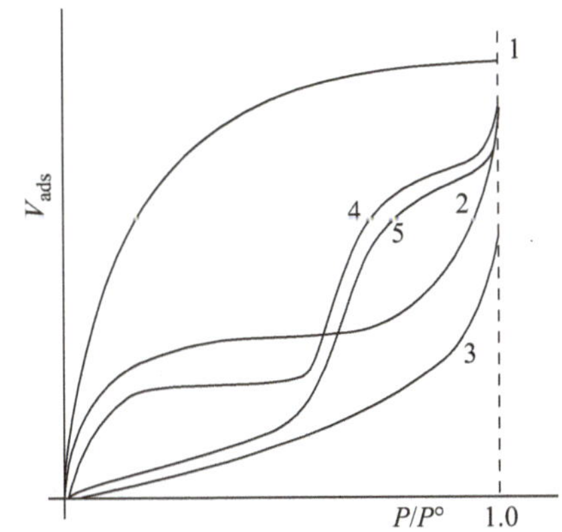

Figure 2.11:
Typical physical adsorption isotherms showing five distinct types: (1) monolayer adsorption only; (2) monolayer followed by multilayer; (3) multilayer without prior monolayer formation; (4) monolayer, then pore filling and finally multilayer; and (5) pore filling and then multilayer without prior monolayer formation.

theory allows a comparison to be made in a much more quantitative way than had hitherto been possible since it allows one to determine, for any solid, a quantity V_m, the so-called monolayer capacity, and hence its (total) surface area.

The BET theory is derived in a similar way to that used for the Langmuir theory outlined above (that a molecule is adsorbed on a free surface site and that a monolayer of such species can be achieved, this having a constant heat of adsorption, ΔH_1) with the extension that additional layers can be adsorbed on top of that layer with heats of adsorption equivalent to the latent heat of vapourization of the gas (ΔH_L); see Figure 2.12. (These subsequent layers also behave according to the Langmuir model.) In most cases, the second and subsequent layers begin to form before the first layer is complete. The resultant equation is:

$$V_{ads} = \frac{V_m cP}{(P_o - P)\{1 + (c - 1)P/P^o\}} \quad (2.25)$$

where V_{ads} is the volume of gas adsorbed, V_m is the volume required to give a monolayer of adsorbed molecules, P is the measured pressure of the gas, P^o is the saturated vapour

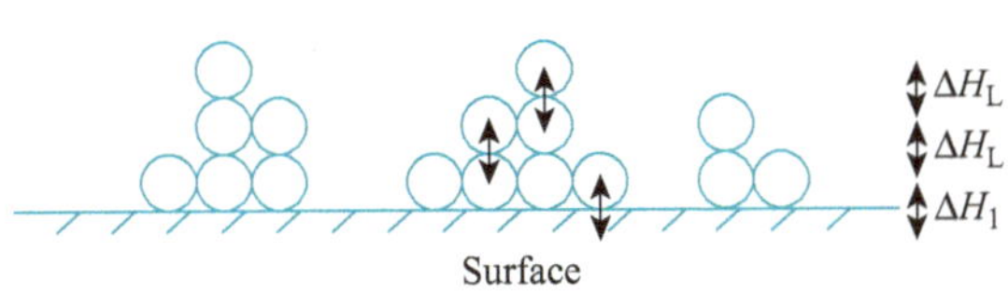

Figure 2.12:
Schematic representation of the formation of first and subsequent layers; note that multilayers start to form before the monolayer is complete.

BOX 2.7 Brunnauer, Emmett and Teller

Stephen Brunnauer, a Hungarian born in Budapest in 1903, was first a student and then a postdoc with Paul Emmett (below). After wartime work on explosives, he spent some years as manager of basic research at Portland Cement before becoming Chair of the Chemistry Department of Clarkson College (see: http://www.caer.uky.edu/energeia/PDF/vol9-1.pdf).

Paul Emmett (1900–1985) became chairman of the Department of Chemical Engineering of John Hopkins University in 1937; in 1943, he joined the Manhattan Project (where he developed a method of enriching U^{235}) and then spent several years and the Mellon Institute of Industrial Research before returning to John Hopkins as W.R. Grace Professor in the Chemistry Department. Following his retirement in 1971, he became a Research Professor at Portland State University (see: http://en.wikipedia.org/wiki/Paul_H._Emmett\).

Edward Teller (1908–2003), also a Hungarian and later a very controversial figure, was then Professor of Physics at George Washington University. He was persuaded to collaborate on the project by Brunnauer. He later became a member of Manhattan Project, becoming known as Father of Hydrogen Bomb. For more details, see the extensive Wikipedia article: http://en.wikipedia.org/wiki/Edward_Teller\.

pressure of the gas at the temperature of adsorption and c is a constant given by the equation:

$$c = \exp\{(\Delta H_1 - \Delta V_L)/RT\} \tag{2.26}$$

Equation 15 can be linearized as:

$$P/\{v(P^o - P)\} = 1/V_{m^c} + (c-1)P/V_{m^c} \times P^o \tag{2.27}$$

Hence, a plot of $P/\{v(P^o - P)\}$ (or $P/P^o/\{v(1-P/P^o)\}$) versus P/P^o will have a y intercept of $1/V_m c$ and a gradient of $(c-1)/V_m c$. The values of the constant c and V_m can hence be determined. (Alternatively, the x intercept is $-1/(c-1)$ from which c can be determined directly.)

Figure 2.13 shows a typical plot of an isotherm which obeys the BET equation. It can be seen that the theoretical curve follows the experimental points quite closely over a range of $0.1 < P/P_o < 0.3$. However, above the latter value, the curves begin to deviate quite significantly from the theoretical ones.

Figure 2.14 shows the equivalent plot of the linearized form of the equation (Eq. 2.26). Again, it can be seen that the experimental values lie close to the theoretical curves up until $P/P^o \approx 0.3$, but that there are significant deviations above that value. Much has been written concerning deviations from the BET equation and attempts continue to be made to develop more accurate equations. Nevertheless, the BET equation is firmly established as a tool to determine the value of V_m and hence the surface area of the adsorbent in question. This is because it allows an unequivocal and reproducible value to be determined for any solid, regardless of the values of ΔH_1 and ΔH_L. The student should beware of applying the equation blindly, for example, by

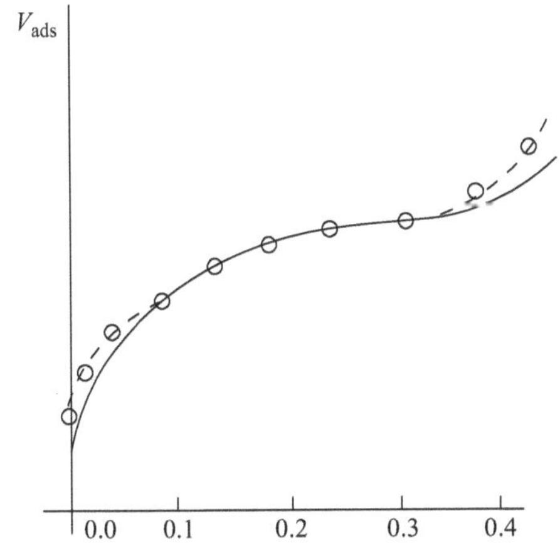

Figure 2.13:
Typical results for the adsorption of nitrogen on a solid surface at 78 K. The solid curve is drawn to fit the BET equation and shows a reasonable fit to the experimental points, especially at $0.1 < P/P_o < 0.3$.

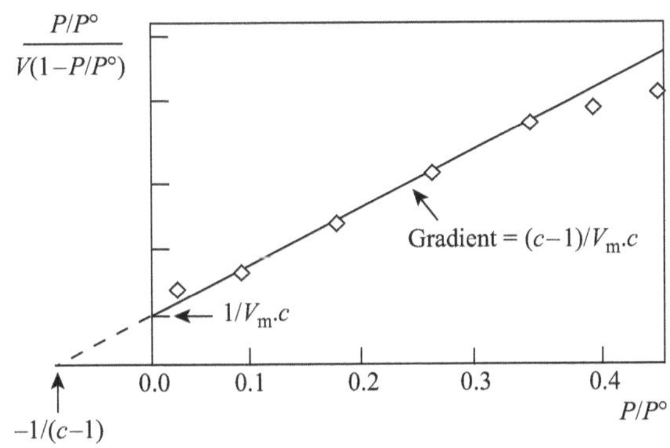

Figure 2.14:
Linearized form of the BET equation showing how V_m can be obtained from the X and Y intercepts or from the gradient and Y intercept.

using computerized best-fit methods, however, as these may wrongly attempt to include points that deviate from the linear reciprocal plots (at lower or higher P/P^o values).

It should also be recognized that the method using nitrogen adsorption at 78 K, the boiling point of liquid nitrogen ($P^o = 1$ Bar, 1014 hPa), works best for relatively high surface areas. This is for the very practical reason that adsorption is most commonly carried out using volumetric or gravimetric methods and the measurements are, therefore, of differences in pressure or weight; if the surface area is too low, the accuracy in determining a pressure or weight difference may be insufficient to give a meaningful surface area. This difficulty using a volumetric system can on occasion be overcome by using a different adsorbent such as Kr or even Xe at 78 K as these atoms have much lower values of P^o: 2 hPa and 4×10^{-3} hPa, respectively. The experiments have to be carried out using equipment capable of measuring accurately pressures in the region of 0.2 hPa and 4 × 10^{-4} hPa respectively, for example, using a McLeod gauge (http://en.wikipedia.org/wiki/McLeod_gauge\); with such equipment, it is possible to measure much smaller uptakes than if nitrogen was used.[10]

Figure 2.15 shows two typical plots obtained for the adsorption of Kr at 78 K on an evaporated metallic film of iron prepared using high vacuum methods before (curve 1) and after (curve 2) exposure to H_2S at 306 K. The corresponding BET plots are shown in Figure 2.16, from which it can be calculated that the V_m values were 69.1 and 55.5 × 10^{-3}cm^3 (stp), respectively.[11]

While visual inspection of the curves of Figure 2.15 allows one to estimate very approximate monolayer values corresponding to the values of "Point B" (B_1 and B_2 shown) of 58.0 and 53.5 × 10^{-3}cm^3 (stp), these values are lower than those obtained by the BET method, particularly for the fresh Fe sample for which the "knee" in the adsorption isotherm is more diffuse. The different shapes of the isotherms are related to different free energies of adsorption of Kr on the fresh and sulfided surfaces, these being related to very different values of the adsorption constant c (401 and 1511, respectively). It, therefore, appears that the Kr is more strongly adsorbed on the sulfided surface than on the un-covered metal surface.

It is generally accepted that the cross-sectional area occupied by a Kr atom under these circumstances is about 0.195 nm^2 and so it can be calculated that the areas of the film before and after reaction with H_2S were 362 and 290 cm^2, respectively. As the area of the surface of

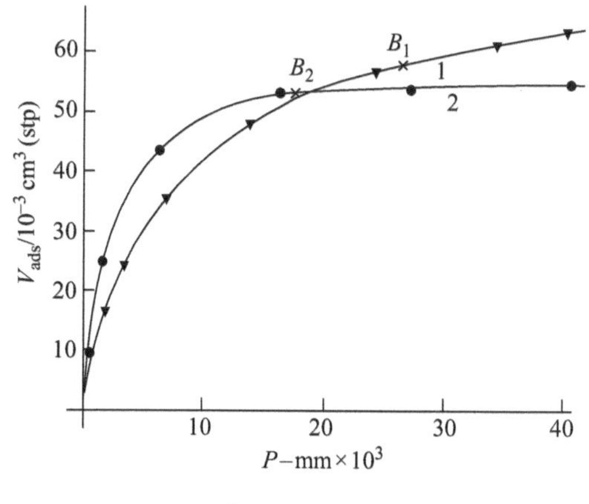

Figure 2.15:
Isotherms for Kr adsorption at 78 K for an evaporated metal film of iron (1) before and (2) after reaction with H_2S at a temperature of 306 K. Source: *J.R.H. Ross, Ph.D. thesis, QUB, 1966.*

[10] M.W. Roberts, J.R.H. Ross, Trans. Faraday Soc., 62 (1966) 2301.

[11] In this example, the amount of gas adsorbed is expressed in units of volume adsorbed; however, numbers of molecules or weights could also be used.

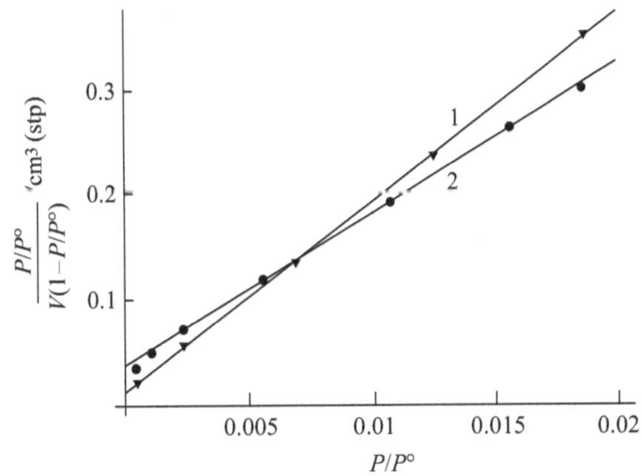

Figure 2.16:
Corresponding BET plots for the data of Figure 2.15.

the pyrex vessel onto which the film had been evaporated was approximately 150 cm^2, the iron film had an area higher than the geometric value.

2.12 Behaviour of Physical Adsorption Isotherms at Values of $P/P^o \geq 0.3$

Figure 2.14 showed that there was significant deviation from the BET equation at values of P/P_0 greater than about 0.3. This deviation is, in part, due to a failure of the BET equation to take into account changes in the adsorption behaviour as the thickness of the adsorbed layer increases. However, a more important contribution to the observed deviation is the onset of so-called pore condensation. It is beyond the scope of this text to examine pore condensation in great detail. Readers interested in expanding their knowledge of the subject should refer to one of the many texts dealing with adsorption and pore condensation.

Every solid exhibiting high surface area consists of small particles, often in the nanometre size range, stacked on one another in either a random or an ordered way.[12] The result is that there will be gaps between the particles and these gaps tend to be relatively uniform in shape and size. They are more often referred to as "pores." This term must not be taken to mean that these pores are always cylindrical or totally uniform in appearance: they can be slit-shaped; they can have the form of ink-bottles, with narrow openings and wider interiors; they can have the form of gaps between platelets; or they can be perfectly uniform as is the case of pores in zeolite crystals (see Figure 2.17). However, for the present purposes, we will assume that all the pores are cylindrical.

When physical adsorption occurs on a solid with cylindrical pores, a monolayer first forms on the external and internal surfaces at relatively low values of P/P_0. As the relative pressure increases, multilayer formation begins to occur on all the surfaces. If we had only external

[12] The preparation of typical catalyst materials is discussed in more detail in Chapter 3.

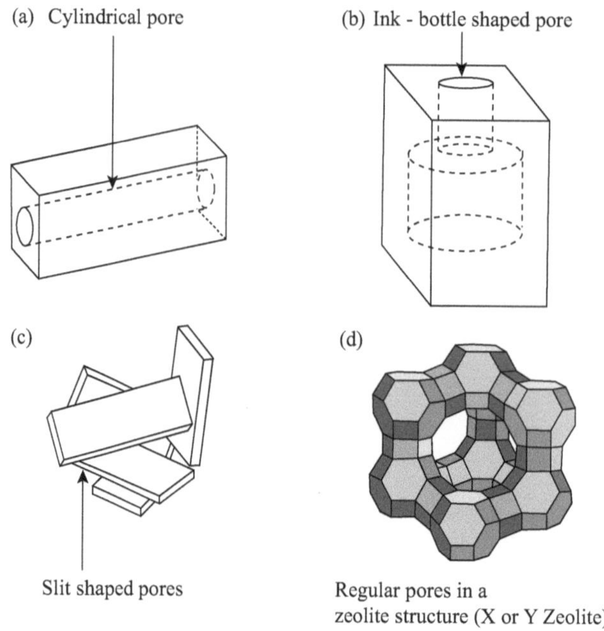

Figure 2.17:
Some typical pore structures: (a) idealized cylindrical pore in a solid matrix; (b) an ink-bottle shaped pore in a solid matrix; (c) slit-shaped pores formed between platelets of a solid; and (d) regular pores in an X or Y zeolite structure.

surfaces, which approximated to flat bodies, we would start to get condensation of a liquid only as P/P_o approached 1.0; in other words, liquefaction occurs at a pressure equal to the saturation vapour pressure of the adsorbate. (Reminder: for nitrogen at 78 K, this occurs at 1 atm. pressure.) Hence, at P/P_o approaching 1.0, the volume adsorbed rises rapidly and tends to ∞. However, if pores are present, they begin to fill (see Figure 2.18) as the pressure is increased at values that are much below P_o, the corresponding value of P/P_o being determined by the pore radius or radii (see below). Once the pores have filled, multilayer adsorption occurs on the external surfaces of the solid as the pressure tends to P_o. If the pressure is then decreased, the isotherm follows the adsorption branch for a short range of P/P_o, corresponding to the removal of the multilayers on the external surfaces. However, the isotherm then continues to a much lower level of P/P_o before desorption from the pores occurs. The "desorption branch" of the isotherm is generally parallel to the "adsorption branch," but this is not necessarily the case.

We will now give a brief explanation of how this adsorption/desorption behaviour occurs. If the solid adsorbent contains uniform cylindrical pores of radius r (see Figure 2.19), the effective saturation vapour pressure at a curved liquid surface within these pores is reduced, and the new value P being given by the Kelvin equation:

$$\text{In } P_o/P' = 2\gamma \text{V}.\cos\varphi/rRT \tag{2.28}$$

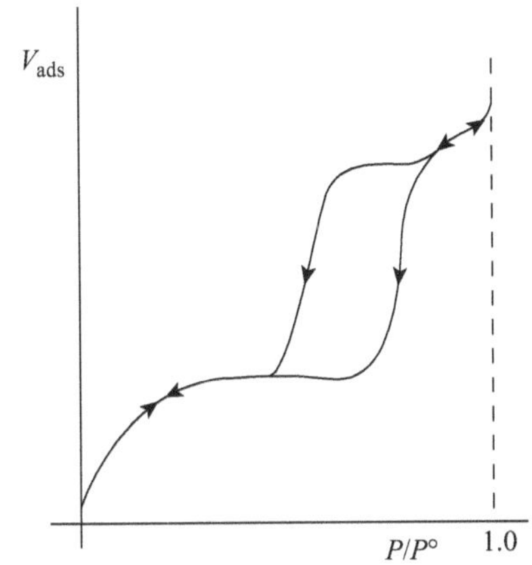

Figure 2.18:
Schematic representation of an isotherm exhibiting hysteresis, showing the filling and emptying of the pores.

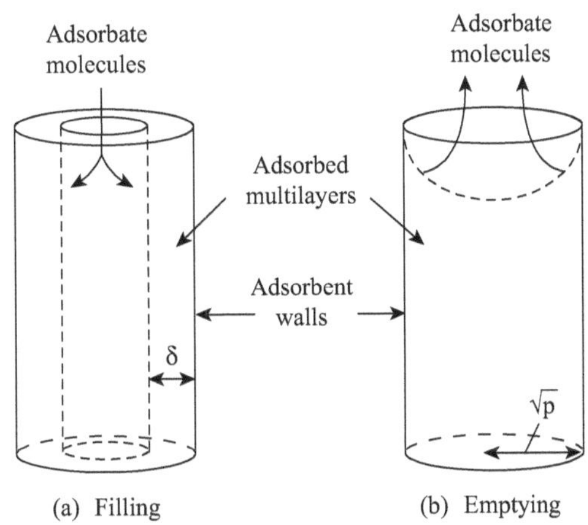

Figure 2.19:
Schematic representation of the filling and emptying of cylindrical pores of radius *r*. The pores fill as shown in (a) by the build-up of molecules in layers within the pores until the pores are filled; however, as they empty (b), desorption now occurs from a spherical interface, which gradually moves into the pores.

where γ is the surface tension of the liquid (8.72 mN/m for N_2), V is its molar volume (34.68 cm^3/mol for N_2), φ is the contact angle of the liquid with the gas (assumed to be zero for liquid N_2) and r is the radius of curvature of the liquid surface. If there are two radii of curvature, r_1 and r_2, then the equation is:

$$\text{In } P_o/P = \gamma \text{V.cos}\varphi(1/r_1/+1/r_2)RT \tag{2.29}$$

When a linear cylindrical pore begins to fill due to condensation within the pore as a result of an increase in the pressure, the condensation occurs on a layer of the adsorbate, which is already present (see Figure 2.19a). The pore surface can be thought of as having two radii, one being $(r - \delta)$, where δ is the thickness of the pore layer, and the other being ∞, corresponding to the linear dimension of the pore. Inserting these values in Eq. 2.28 gives:

$$\text{In } P_o/P' = \gamma V\{1/(r\text{-}\delta)\}RT \quad (2.30)$$

If the pores are entirely uniform, the volume adsorbed increases rapidly at a single value of P' until the pores are completely filled; if there is a range of pore sizes, the volume adsorbed increases less sharply as the pressure is increased. The data can be used to determine the pore diameter (or the pore size distribution if all the pores are not the same) as long as the thickness δ is known; the value of δ is calculated from the amount adsorbed at the value of P/P_o at the onset of capillary condensation.

When the pores are completely filled, all that remains to be covered is the external surface and there may be a final corresponding slight increase in volume adsorbed until the vessel containing the solid is completely filled. The total volume adsorbed up to and including that last section of the isotherm is equivalent to the total volume of the pores in the solid, an important parameter in relation to catalysis.

If the saturated surface is then evacuated in a controlled way (in other words, the adsorbed gas is removed in known steps), a desorption branch to the isotherm may also be obtained (see Figure 2.18). Desorption now occurs from the pores in a different way to the way in which adsorption occurred: the adsorbate is desorbed from hemispherical surfaces in each pore. The equation which now applies is Eq. 2.29 (i.e. both r_1 and r_2 are equal to the actual pore radius, r, without adsorbed material – see Figure 2.19b). As shown in Figure 2.18, the "desorption branch" of the isotherm is different from the adsorption branch. Values of r calculated from both branches should agree but that is not always the case.

TASK 2.6 Surface Area and Pore Structure

For a good starting point to study the scientific literature on adsorption methods to determine surface area and pore structure, see a relatively recent article by K.S.W. Sing in Adv.Colloid Interface Sci. 76 (1998) 3. Using Scopus or Web of Science, examine other more recent articles citing Sing's review and tabulate the most interesting of these.

A full discussion of pore filling and the shapes of pore systems and of more advanced treatments of adsorption data (e.g., the BJH method[13]) is beyond the scope of this book. It is

[13] E.P. Barrett, L.G. Joyner, P.P. Halenda, "The determination of pore volume and area distribution in porous substances. I. Computations from nitrogen isotherms," J. Am. Chem. Soc., 73 (1951) 373.

recommended that the interested reader should study the subject in more detail using one of the standard texts on the subject. One of the earliest books, a classic, was *Adsorption, Surface Area and Porosity* by S.J. Gregg and K.S.W Sing (2nd Edition), Academic Press, New York and London (1982). A recent book is *Characterization of Porous Solids and Powders: Surface Area, Porosity and Density*, by S. Lowell, J.E. Shields, M.A. Thomas and M. Thommes, Kluwer Academic Publishers (2004).

CHAPTER 3

How Does a Catalyst Work?

Chapter Outline

3.1 Introduction

This chapter will concentrate mainly on the catalyst: how does one make an effective catalyst that will not only catalyse the desired reaction, but will also have a life-time long enough to survive the stringent conditions which prevail in a catalytic reactor; and how does one characterize such a catalyst and study all the reactions that take place on its surface during reaction? In Chapter 1, we discussed the nature of a catalyst and traced the history of the development of the applications as well as the understanding of catalysis, while in Chapter 2, we examined the phenomena of adsorption, chemical and physical, which are so important in the functioning of a catalyst. We will now draw the content of these two chapters together, showing how adsorption is an essential step in the operation of a catalyst and how it determines the behaviour of a catalyst under operation conditions. We will then consider some typical catalyst formulations and some of the common catalytic reactions occurring over metals, oxides and sulfides.

Heterogeneous Catalysis. DOI: 10.1016/B978-0-444-53363-0.10003-9

3.2 The Catalytic Process

3.2.1 Bimolecular Processes

In Chapter 1, we considered the abstract homogeneous reaction:

$$A + B \leftrightarrows C + D \tag{3.1}$$

As written, this equation represents an equilibrium situation in which species A reacts with species B reversibly to give species C and D. The equilibrium position for the reaction is determined by thermodynamics and, once equilibrium is achieved, the rates of the forward and reverse reactions are equal. The energetics of the process are shown schematically in Figure 3.1.

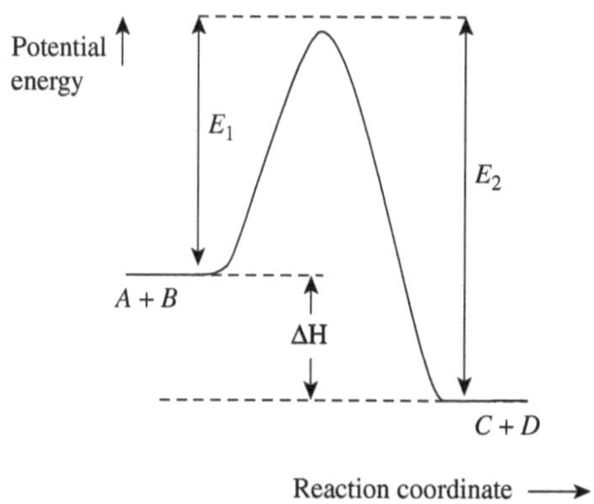

Figure 3.1:
Energetics of the reaction: $A + B \rightarrow C + D$

In this figure, the reaction is shown as being exothermic (ΔH −ve) but it could equally well have been shown as being endothermic. In order to transform into the products C and D, the reactants A and B must come together to form an activated (or transition state) complex $A{-}B^{\ddagger}$ which then decomposes to form the products.[1] The activation energy for the formation of the complex is E_1 and this has a numerical value well above the average thermal energy of the molecules A and B. (If the reaction occurs in the gas phase, the acquisition of the activation energy is most easily visualized by considering that the molecules pick up the energy by a series of gas-phase collisions.) Once C and D are formed, they can transfer back to A and B by the reverse process, the activation energy for that change being E_2. When the rate of the reverse reaction r_2 becomes equal to that of the forward reaction r_1, the all-over reaction has reached a dynamic equilibrium. If the forward and reverse reactions are both elementary processes,

[1] The formation of transition state complexes is handled in standard textbooks on reaction kinetics. The subject has also been extended to consider reactions at surfaces, see, for example, S. Glasstone, K.J. Laidler, H. Eyring, The Theory of Rate Processes McGraw-Hill, New York, 1941, but the subject will not be considered further here.

$$r_1 = k_1[A][B] \quad \text{and} \quad r_2 = k_2[C][D] \tag{3.2}$$

from which we can derive:

$$r_1/r_2 = k_1/k_2\ [C][D]/[A][B] = k_{eq} \tag{3.3}$$

TASK 3.1

Show that $E_1 - E_2 = \Delta H$. Remember that $k_1 = A\ \exp\text{-}E_1/RT$ (This is the Arrhenius equation in which A is the Arrhenius constant and E_1 is the activation energy.)

We now consider the situation when A and B are reacted in the presence of a solid surface which will adsorb both A and B. As shown in Eq. (3.4), the adsorbed species A_{ads} and B_{ads} may then transform on the surface to give C_{ads} and D_{ads} and these can in turn desorb to give the gaseous products C and D:

$$A + B \xrightarrow{1} A_{ads} + B_{ads} \xrightarrow{2} (AB)^{\neq}_{ads} \xrightarrow{3} C_{ads} + D_{ads} \xrightarrow{4} C + D \tag{3.4}$$

Reaction (3.4) on the surface will occur in preference to the gas phase bi-molecular process[2] as long as rates of the constituent steps (adsorption, surface reaction and desorption) are all more rapid than the all-over rate of the gas phase process. The situation can be represented schematically as in Figure 3.2, in which, for simplification, single energy troughs are shown for A_{ads} and B_{ads} and for C_{ads} and D_{ads} and a single step is shown for the inter-conversion to the adsorbed products.[3] Two curves are shown in the figure. Curve (i) represents the situation in which both A and B adsorb without any activation energy for the adsorption step while curve (ii) represents a situation in which there is an activation energy for the adsorption of A and B. Once A and B are adsorbed, they can transfer, via an adsorbed complex $(AB)^{\neq}{}_{ads}$, to give adsorbed C and D and these surface species then desorb to give products.[4] The most important message to get

[2] Collision theory explains the acquisition of the activation energy required for the gas-phase process by considering that it occurs via random gas-phase collisions between A and B. Once the products are formed, the excess energy is dissipated by further gas-phase collisions.

[3] This representation is unrealistic as there will be different heats of adsorption for A and B as well as for C and D; in addition, it is probable that a number of steps will be involved in the surface transition. Further, if there are activation energies for the adsorption for the two reactants, these will be different. The same will be the case for the desorption process. The diagram should be seen only as a means of gaining a picture of how the catalysed reaction occurs.

[4] Many other possible sets of curves could be imagined. For example, the adsorption of A could have a high activation energy, while the adsorption of B might be un-activated; alternatively, the desorption of C or D might have a high activation energy. Furthermore, as A and B may adsorb preferentially on sites that are some distance apart, there might also be activated diffusion of A and/or B over the catalyst surface to allow them to reach sites at which reaction can occur.

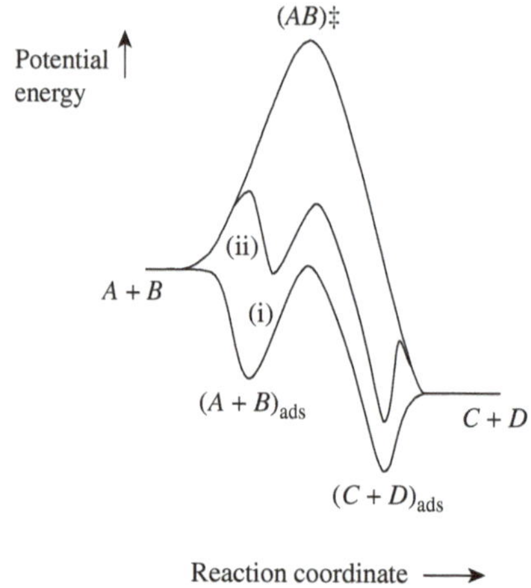

Figure 3.2:
Schematic representation of the energetics of the catalysed reaction: $A + B \rightarrow C + D$

from a very schematic diagram of this type is that the catalyst surface provides an alternative route for the reaction and that this route is energetically much more favoured than is the gas phase reaction; in the schematic representation shown in Figure 3.2,

BOX 3.1 Hydrogen Exchange Reactions

The early development of the theories of bimolecular reactions on catalyst surfaces were based on reactions of hydrogen species: ortho (o-) to para (p-) hydrogen exchange and H_2/D_2 exchange. Ortho hydrogen is a molecule of hydrogen in which the proton spins are parallel (i.e. in the same direction), while para hydrogen has spins that are antiparallel. The para modification is the favoured one at low temperatures (ca. 80 K) and the transfer to this form may be catalysed by graphite. Its transfer back to the para modification has been examined over a number of different catalysts, for example, evaporated films of transition metals, and it has been shown that the reaction generally occurs in a sequence involving the dissociative adsorption of the hydrogen on the surface followed by recombination with desorption:

$$\text{o-}H_2 \rightarrow 2H_{ads} \rightarrow \text{p-}H_2$$

A similar mechanism applies to the hydrogen–deuterium exchange reaction:

$$H_2 + D_2 \rightarrow 2H_{ads} + 2D_{ads} \rightarrow 2HD$$

While the o- to p-H_2 exchange reaction required the use of difficult thermal conductivity measurements to distinguish between the two forms, it is simple to study the hydrogen–deuterium reaction using mass-spectrometry. Examination of these two reactions led to the development of two well-established mechanisms of surface reaction: the Langmuir–Hinshelwood and Eley–Rideal mechanisms. We will return to a discussion of these in Chapter 6.

the surface reaction is likely to be rate determining and the apparent activation energy for the all-over reaction is then associated with the barrier for the surface reaction. The second message to gain from the diagram and discussion is that the catalysed reaction may have any number of steps, any one (or combination) of which may be rate determining.

3.2.2 Unimolecular Processes

Similarly to a bimolecular process, a unimolecular decomposition reaction can also be catalysed by a surface. Figure 3.3 shows the energetics of a simple model reaction, the decomposition of molecule A to give products.

The upper curve in Figure 3.3 represents the uncatalysed reaction.[5] As with the bimolecular reaction discussed above (Figures 3.1 and 3.2), the reaction is taken to be exothermic, but the arguments apply equally to an endothermic reaction. For the reaction in the gas phase, the activation energy for the forward reaction is shown as E_1 and that for the reverse reaction is

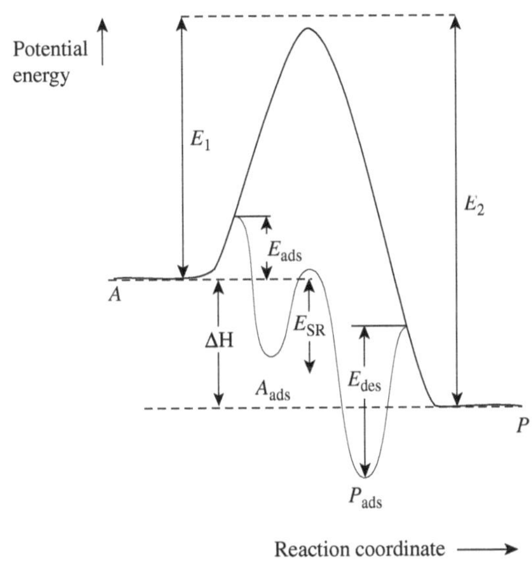

Figure 3.3:
Schematic diagram of the energy profile for a reaction $A \rightarrow B$ without and with a catalyst.

[5] For a gas-phase unimolecular reaction, Lindemann developed a modified collision theory to explain how the activation energy for the reaction is obtained. This is based on collisions between the reactant A and inert molecules (M) or unreacting molecules of A. In these collisions, A acquires energy by collisions with M or A with the result that an activated molecule A≠ is formed; A≠ and the product molecules then lose their excess energy by transfer to M or to the other molecules. For a surface reaction, the energy could easily be picked up from the surface.

shown as E_2. The enthalpy of the reaction ΔH is given by $E_1 - E_2$; in the figure as drawn, as E_2 is greater than E_1, the value of ΔH is negative.

When the reactant can adsorb on a surface, a new state is formed, A_{ads}, with a heat of adsorption, $\Delta H_{A(ads)}$. Depending on the system, there may or may not be an activation energy E_{ads} for this adsorption step. Once the species A is adsorbed, it may be transformed into P_{ads} by a surface reaction process with an activation energy E_{SR}. The product species P is shown as being adsorbed with an adsorption energy of $\Delta H_{P_{ads}}$. Finally, P is desorbed with an activation energy E_{des}. The initial and final states for the uncatalysed and catalysed reactions are exactly the same: all that the catalyst has done is to change the all-over energetics, providing an alternative route from reactant to product. This is a fundamental tenet of catalysis: a catalyst can bring about a change in the rate of a chemical reaction, but has no effect on the all-over thermodynamics of the reaction, that is, the equilibrium position.

BOX 3.2 Decomposition of Formic Acid

An example of such a reaction is the formic acid decomposition reaction (5):

$$HCOOH \rightarrow H_2 + CO_2 \tag{5}$$

which is slightly exothermic, having a value of ΔH° of $-$ 32 kJ mol^{-1}.

As there are two discrete product species, the representation shown in Figure 3.3 is again (as for the bimolecular reaction discussed above) not strictly correct as the desorption of each species should be shown. There is a further complication: alternative products, CO and H_2O, could be formed and so the "selectivity" of the catalyst becomes important (see below). Products formed in the reaction depend on the ease of formation of different surface species and the ease with which they can transform into others.

3.2.3 Reversibility in Catalytic Processes

An important tenet of catalysis is illustrated by examination of Figure 3.3: that the catalyst for the forward reaction will also, under suitable conditions, catalyse the reverse reaction and that the step or steps which dictate the rate for the forward reaction will also dictate the rate for the reverse reaction. (Similar arguments can also be applied to the bimolecular reactions discussed above.) For example, if it is found that the surface reaction:

$$A_{ads} \rightarrow P_{ads} \tag{3.5}$$

is the rate determining step (rds) for the forward reaction, then the reverse surface reaction:

$$P_{ads} \rightarrow A_{ads} \tag{3.6}$$

is the rate determining step for the reaction in the other direction.

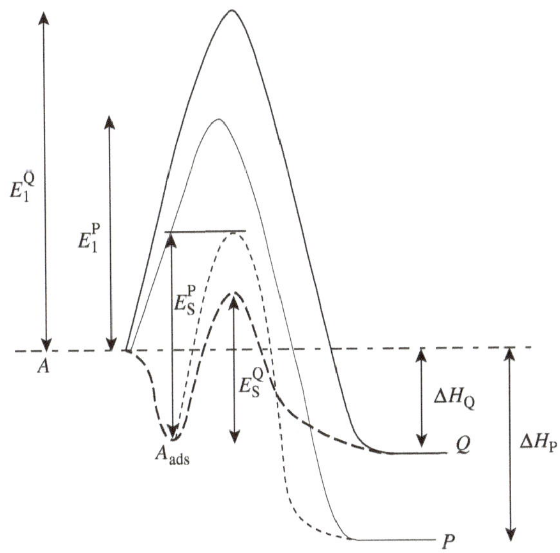

Figure 3.4:
Schematic representation of the energetics of two parallel reactions A → P and A → Q, showing that a catalyst can change the selectivity.

3.2.4 Selectivity in Catalytic Processes

Another essential aspect of catalysis can be illustrated by an extension of the example shown in Figure 3.3. If A can react to give two different products, *P* and *Q*, the presence of the catalyst can have a very significant effect on the "selectivity" of the reaction as shown in Figure 3.4. Consider first the uncatalysed reaction. *A* can now give the two products, *P* and *Q*, with rates determined by the activation energies $E_1{}^P$ and $E_1{}^Q$; in other words, as long as there is no significant difference in the Arrhenius constants for the two reactions, *P* will be formed more rapidly than *Q* and *P* will therefore be the product that is formed selectively. (In Figure 3.4, the lower activation energy is associated with the reaction to give the product (*P*) for which the enthalpy of formation is more negative, but that is not necessarily the case.) If a catalyst is now introduced on which *A* adsorbs to give A_{ads} and the latter can, in turn, decompose to give either *P* or *Q*, the relative values of the activation energies for the two processes may be reversed. As long as the pre-exponential factors do not differ greatly, the molecule *Q* in the example shown in Figure 3.4 will now be the selective product even though *P* may be thermodynamically favoured.[6] It is the ability of a catalyst to affect the selectivity of a reaction, often producing products that are not found in the equivalent non-catalytic systems, that makes catalysis so important in practice. An

[6] The diagram is simplified by omitting troughs for adsorbed P and Q; if adsorption of the products was significant, other steps such as desorption of products or even rearrangement of A_{ads} might have to be considered. These could even be "rate determining".

important example is the selective oxidation of hydrocarbons. Non-catalytic oxidation reactions typically produce only the two most thermodynamically favoured products, CO_2 and H_2O; however, selective catalytic systems are known which can produce molecules such as the corresponding aldehydes and/or ketones, by selective insertion of oxygen, or olefins, by selective removal of hydrogen ("oxidative dehydrogenation"). We will return to the subject of catalytic oxidation in much more detail in Chapter 8 (Section 4).

3.2.5 Kinetics and Mechanism

In outlining the ways in which adsorbed molecules react on a catalyst surface to form products, we have concentrated on the energetics of the all-over catalytic reaction and on how the presence of the surface affects the reaction, sometimes even changing the product formed. Energetics and kinetics are inevitably closely linked and we have already mentioned the importance of the activation energy barriers for the various steps in a catalytic process. However, we will reserve a full discussion of the most commonly encountered types of kinetic expressions for catalytic processes to Chapter 6 in which we will discuss topics such as Langmuir–Hinshelwood, Eley–Rideal and Mars–van Krevelen mechanisms. Before doing that, we will discuss in this chapter the different types of catalyst in common use; we will then move on, in Chapter 4, to discuss the important topic of catalyst preparation, giving examples of some of the most commonly used approaches to the preparation of practical catalysts, discussing also the characterization of such catalysts, and will then, in Chapter 5, discuss the problems of mass and heat transfer in catalytic processes.

3.3 The Catalyst and the Catalytic Site

So far, apart from mentioning the compositions of the catalysts used in various processes discussed from an historical perspective in Chapter 1 and the phenomenon of adsorption in Chapter 2, we have discussed only briefly the nature of the catalyst or the catalytic site. We will now give a brief general description of the most common types of catalyst currently in use, particular attention being given to the nature of the surfaces of these materials. We will return in detail to the topics of catalyst preparation and characterization in Chapters 4 and 5.

There are many different types of catalysts in use in various processes. However, most commonly encountered as catalysts are metals, oxides and sulfides. Although each may exist as single component materials, the active material is often accompanied by other components, e.g. as "support" or "promoter"; further, a catalyst may have several different active components and, for example, one of these may be a metal and the other an oxide. It is important to recognize that the active form of a catalyst can be highly dependent on the reaction mixture to which it is exposed: whether the conditions are reducing or oxidizing or even sulfiding (see Box 3.3). We will now discuss briefly some examples of active metals, oxides and sulfides and their combinations.

BOX 3.3 Oxidizing Versus Reducing Conditions

Let us consider first the complete oxidation of a simple molecule such as methane:

$$CH_4 + O_2 \rightarrow CO_2 + 2H_2O$$

This reaction can be catalysed by most metals. However, whether or not the operating catalyst surface will actually be in the metallic or oxidic form will depend on the thermodynamics of the oxidation of the metal M:

$$M + 1/2\,O_2 \rightleftarrows MO$$

If this reaction is considered an equilibrium, the equilibrium constant, K_p is given by:

$$K_p = a_{MO} / \left(a_M \cdot P_{O_2}^{e^{1/2}} \right)$$

where a_M and a_{MO} are the activities of M and MO.

Since the activities of both solids are unity,

$$K_p = P_{O_2}^{e^{-1/2}}$$

Hence, the value of the so-called equilibrium pressure $P_{O_2}^{e}$ determines whether or not a metal will be oxidized in the presence of a given partial pressure of oxygen, P_{O_2}. If the partial pressure of oxygen in the reaction mixture, P_{O_2}, is above the equilibrium pressure, $P_{O_2}^{e}$, then the metal will be oxidized; however, if the partial pressure is below the equilibrium pressure, the metal will remain in its reduced form. In practice, this means that noble metals such as Pt, Ir or Au remain in their metallic states while operating under the conditions used for methane oxidation. However, Pd lies on the thermodynamic borderline and there is strong evidence that the active phase for methane oxidation is the oxide.

If a metal M is to be used for the steam reforming of methane:

$$CH_4 + H_2O \rightarrow CO + 3H_2 \tag{3.7}$$

the state of oxidation of the metal will depend on the equilibrium:

$$M + H_2O \rightleftarrows MO + H_2 \tag{3.8}$$

for which the (temperature dependent) equilibrium constant K_p is given by:

$$K_p = P_{H_2} / P_{H_2O} \tag{3.9}$$

The most commonly used metal for steam reforming is nickel (see Section 3.4, also Chapters 4 and 8) for which the thermodynamics are such that the metallic state is favoured at most temperatures as long as the partial pressure of hydrogen in the mixture is above about 1% of the total mixture. In practice (see later), the catalyst will be pre-reduced in a hydrogen-rich gas and is maintained in its reduced state by the addition of small quantities of hydrogen to the reactor feed (or by recirculation of some of the product of the reaction).

The same types of arguments apply to many other situations in which changes of the catalyst surface are possible under reaction conditions, for example, the gradual "sulfidation" of a metal oxide under conditions where H_2S is present in the reaction mixture:

$$MO + H_2S \rightleftarrows MS + H_2O \tag{3.10}$$

While such sulfidation is sometimes desirable,[7] irreversible poisoning of a metal or metal oxide by trace quantities of H_2S can also occur.

While the use of bulk thermodynamic quantities can frequently give adequate information about possible formation of new phases, it has to be recognized that the quantities involved in the thermodynamics of surface and bulk processes are not necessarily the same: they will differ due to the fact that a metal atom in the surface of a catalyst will not have exactly the same environment as that in the bulk of the same metal.

3.4 Catalysis by Metals

3.4.1 Introductory Remarks

Table 3.1 lists some examples of metallic catalysts used in commercial processes. Although nearly all metals in the periodic table can be used as catalysts, those most commonly encountered in common catalyst formulations are the transition metals since these tend to be more reactive as a result of their unfilled d-bands.[8]

While historically much attention was given to attempts to correlate the nature of the unfilled d-bands to the catalytic properties of metals, much more attention is now given to the nature of the chemical bonding between the metal and the reactant species (two-dimensional chemistry). Similar remarks apply to catalysis by oxides, sulfides, etc.

A number of catalytic processes can be carried out over pure metals. However, metals are most commonly "supported" on oxides and more; furthermore, more examples are being reported in which the oxide is not just an inert "support" but instead contributes to the all-over activity and selectivity of the catalyst. We will in later chapters encounter many examples of the importance of the support used. In most of these, the metal surface area is the most important parameter since the rate of any catalytic reaction is under most circumstances proportional to the number of metal atoms exposed. However, there are exceptions; for example, the catalytic properties of gold catalysts are very dependent on the way in which the catalysts are made and are often only active if gold nanoparticles are present on the support (see Box 3.4).

---

7 The $CoO/MoO_3/Al_2O_3$ catalysts used for hydrodesulfurization is sulfided in its active form. The sulfidation is done using H_2S or an organo-sulfur compound.

8 The d-band theory of catalysis is discussed in many of the earlier standard textbooks. See, for example, Bond, Heterogeneous Catalysis, Principles and Applications, Oxford Science Publications, 1987 (p. 40).

Table 3.1: Some catalysts used in commercial processes.

Process	Catalyst	Comments
Ammonia synthesis	Promoted iron	Contains K and other promoters
Ammonia oxidation	Pt/Rh gauze	Gauze becomes severely pitted during use
Fischer–Tropsch synthesis	Fe or Co on support	Support originally keiselguhr, includes various promoters such as K
Steam reforming of methane	Ni on support (typically Al_2O_3)	Variety of supports and promoters used
Methanol synthesis	$Cu/ZnO/Al_2O_3$	
Methanol oxidation	Unsupported silver BiMo??	Ag uses excess methanol; Bi/Mo uses excess air
Reforming of hydrocarbons	Pt/Re/support	
Automobile exhaust treatment	$Pt/Pd/CeO_2/Al_2O_3$	Prepared by wash-coating monolith support
Selective reduction of NOx in flue-gas	V_2O_5/TiO_2/matrix support	Reducing agent commonly ammonia or urea
Ethylene oxidation	$Ag/\alpha\text{-}Al_2O_3$	
Fat hardening	Ni/Al_2O_3	Pre-reduced catalyst stored in vegetable oil

BOX 3.4 Catalysis by Gold

Although it was shown by Trapnell (see Table 2.2) that gold would adsorb molecules such as O_2, CO, C_2H_2 and C_2H_4, gold has long been considered to be relatively inactive as a catalyst. However, two groups of researchers led by M. Haruta[a] and G.J. Hutchings[b] have relatively recently shown that carefully prepared gold nanoparticles supported on a wide variety of supports (particularly the oxides of the first row of the transition series such as Fe_2O_3) are active for various reactions including CO oxidation. An early application of these catalysts pioneered by Haruta and his colleagues was the use of such catalysts for the room temperature destruction of odours in lavatories. The field of catalysis by gold has now become a hot topic and a series of conferences discussing the topic has been instituted.[c] A recent book by Bond, Louis and Thompson gives a very good overview of the subject. They point out that although small gold particles can be made on almost any type of support, the conditions of preparation and storage are critical to avoid gold particle sintering.

---

[a] M. Haruta, Catal. Today 36 (1997) 393; M. Haruta, Gold Bullet., 37 (2004) 27.

[b] B. Nkosi, N.J. Colville, G.J. Hutchings, J. Chem. Soc., Chem. Commun. (1988) 71; G.J. Hutchings, Gold Bull., 29 (1996) 123.

[c] The most recent of these international meetings was held at the University of Heidelberg in 2009 (see Gold Bulletin, 42 (2009), issues 1 and 2 (http://www.goldbulletin.org); the 2006 meeting was held in the University of Limerick (Catalysis Today, 122, issue 3/4).

TASK 3.2

Using references a and b of Box 3.4, examine the methods of making stable and active gold catalysts. What industrial applications of gold catalysts are currently available or proposed?

Many metal-catalysed reactions are ones that are carried out in their entirety under reducing conditions, the best known of these being processes such as ammonia synthesis or paraffin isomerization (Table 3.1) in which neither oxygen nor sulfur is a component of the reaction mixtures. Under these conditions, all the metals will be thermodynamically stable and many of them are potential candidates as catalysts. However, other parameters are also important: will the metal be susceptible to sintering as a result of the diffusion of surface metal atoms under reaction conditions (low Tamman temperature); will the metal be susceptible to other reactions such as carbon deposition; will it catalyse undesired side reactions. If the reaction mixture contains oxygen or sulfur, then the situation is different and the metal must be chosen from a more limited group that is not susceptible to oxidation or sulfidation under the reaction conditions to be used.

3.4.2 Unsupported Metal Catalysts

Most metallic catalysts consist of metal particles supported on a high area oxide material in order to provide as high a metallic surface area as possible. However, there are several examples of the use of unsupported metals. One well-known example of the latter type (see also Section 1.2.4) is the use of finely woven Pt/Rh gauzes for the oxidation of ammonia:

$$4NH_3 + 7O_2 \rightarrow 4NO_2 + 6H_2O \tag{3.11}$$

This reaction occurs in two steps:

$$4NH_3 + 5O_2 \rightarrow 4NO + 6H_2O \tag{3.12}$$

followed by:

$$2NO_3 + O_2 \rightarrow 2NO_2 \tag{3.13}$$

The NO_2 is then absorbed by water to form nitric acid:

$$3NO_2 + H_2O \rightarrow 2HNO_3 + NO \tag{3.14}$$

The gauzes used gradually change in form during use as result of surface roughening until the gauzes break and have to be replaced.[9] The reason for this gradual change is not fully known but its progress can be slowed by inclusion of about 10% Rh in the gauze material.[10]

The ammonia oxidation reaction is carried out in a very short contact time ($\leq$10-3 s) at temperatures above 1000 K and the rate of reaction is controlled by the rate of mass transfer of the reactants to the surface of the gauze; as the all-over reaction is highly exothermic, the process is self-sustaining once it has started.

---

9 See, for example, F. Sperner, W. Hohmann, Platinum Metals Rev. 20 (1976) 12; free access to all issues of this journal is available at http://www.platinummetalsreview.com/dynamic/advancedsearch.

10 Some unreported experiments carried out in our laboratory in about 1968 showed that small amounts of a volatile Pt oxide (probably PtO) could be detected mass-spectrometrically in the gas phase directly above the catalyst when a Pt filament was heated in a very low pressure of oxygen at a temperature of around 1273 K.

Pt gauze materials have also been used for the catalytic removal of volatile organic compounds (VOCs) from gas mixtures by means of total oxidation. Another use of an unsupported metal is that of unsupported silver (electrochemically produced granules arranged in a thin bed about 1 cm in depth) for the selective oxidation of methanol to give formaldehyde:

$$CH_3OH + 1/2\,O_2 \rightarrow HCHO + H_2O \tag{3.15}$$

About half of the formaldehyde produced worldwide comes from the silver process. This reaction, first observed by August Wilhelm von Hofmann, uses an excess of methanol and hence some dehydrogenation also occurs in parallel:

$$CH_3OH \rightarrow HCHO + H_2 \tag{3.16}$$

The other process used for formaldehyde production employs excess oxygen with Fe/Mo/O or Fe/V/O catalysts (FORMOX process; http://en.wikipedia.org/wiki/Formaldehyde).

As unsupported metallic catalysts are often susceptible to sintering due to relatively high temperatures of use, the metal particles have to be stabilized in order to give as high a surface area as possible. As sintering often occurs by a surface transport mechanism, this stabilization may be achieved by the addition of a "promoter" that helps to anchor the metal atoms on the surface and prevent surface migration and/or particle coalescence. In the case of the platinum gauzes used for ammonia oxidation referred to above, the addition of Rh slows down surface rearrangement. Another example of stabilization of metal particles is found in the case of Raney nickel (see Box 3.5), the skeletal structure of which is stabilized by small quantities of alumina on the surface of the Ni lattice.

BOX 3.5 Stabilization of Ni Particles Used in Fat Hydrogenation – Raney Nickel

At the beginning of the twentieth century, unsupported nickel powders were used for the hydrogenation of vegetable oils and these suffered from serious sintering problems. In 1926, the American scientist and engineer Murray Raney discovered that it was possible to produce a more stable Ni-based catalyst from Ni/Al alloys by leaching out the majority of the aluminium using concentrated sodium hydroxide, this leaving a "skeletal" Ni structure which has a high activity for the hydrogenation reaction. According to a Wikipedia article discussing Raney nickel (http://en.wikipedia.org/wiki/Raney_nickel), the activated material contains both Ni metal and a Ni/Al alloy that helps to stabilize the Ni particles. (An alternative explanation, see below, is that some alumina particles are formed, and that these help to stabilize Ni.) These Raney Ni catalysts (Cu, Co and Ru equivalents can also be made) are not only used for fat hydrogenation but also for hydrogenation of nitrocompounds and other similar reactions. Raney Ni has also been used commercially for the production of bipyridyl, a precursor used in the production of pesticides, by the combination of two molecules of pyridine (http://en.wikipedia.org/wiki/2,2'-Bipyridine):

$$2C_5H_5N \rightarrow (C_5H_5N)_2$$

(Continued)

BOX 3.5 Stabilization of Ni Particles Used in Fat Hydrogenation – Raney Nickel—*Cont'd*

However, there is no particular advantage of using Raney nickel catalysts for this purpose as better conversions, accompanied by higher catalyst stability, can now be obtained using pre-reduced coprecipitated Ni/Al catalysts.

The particular properties of Raney nickel which make it superior to unsupported finely divided Ni have been associated with the "skeletal" structure formed during the leaching process. Whether this skeletal structure is associated with a residual Ni/Al phase or with the presence of an amorphous alumina phase, the Ni particles are more resistant to sintering than if only metallic nickel was present. This is probably due to the residual Ni/Al phase or the amorphous alumina present sitting between Ni particles and preventing them from coming into physical contact with one another.[10] A similar model has been put forward for the stabilization of coprecipitated Ni-Al_2O_3 catalysts by an alumina phase present between (larger) Ni particles.*

* See L.E. Alzamora, J.R.H. Ross, E.C. Kruissink, L.L. van Reijen, Coprecipitated nickel-alumina catalysts for methanation at high temperatures, Part 2, Variation of total and metallic areas as a function of sample composition and method of pretreatment, J. Chem. Soc., Faraday Trans., I 77 (1981) 665–681.

3.4.3 Supported Metal Catalysts

In many of the examples of unsupported metal catalysts discussed above, the metallic surface area is relatively unimportant as the rate of the reaction for which these metals is used is frequently dictated by the external metal area, the rate of reaction being related to the rate of supply of reactant to the external surface of the metal (e.g. in ammonia oxidation, the rate of supply of the ammonia to the Pt surface by gaseous diffusion is rate determining; see Chapter 7). For reactions in which the rate is determined by the rate of surface reaction, the metallic area of the catalyst becomes the most important factor. Hence, it is important to make and stabilise the highest possible and easily accessible metal surface area. Although finely divided metallic powders can be made with high surface areas, these are very unstable under reaction conditions. Supported metal catalysts are therefore used.

As the name implies, a supported metal is one in which the metal crystallites exist on the surface of a support. Methods have been developed (see Chapter 4) to prepare very small metal crystallites on high-area supports, the particles having diameter values of the order of nanometers.[11] The most commonly used supports are single oxides such as alumina or silica, complex oxides such as silica-alumina or zeolites and active carbons. We will return in

[11] Many of the methods used for the preparation of these nanosized particles were developed long before the term nanotechnology was introduced. Nanotechnology can be defined as "the study of controlling matter on an atomic and molecular scale" (en.wikipedia.org/wiki/Nanotechnology) and so the preparation of supported metal catalysts clearly falls within this definition.

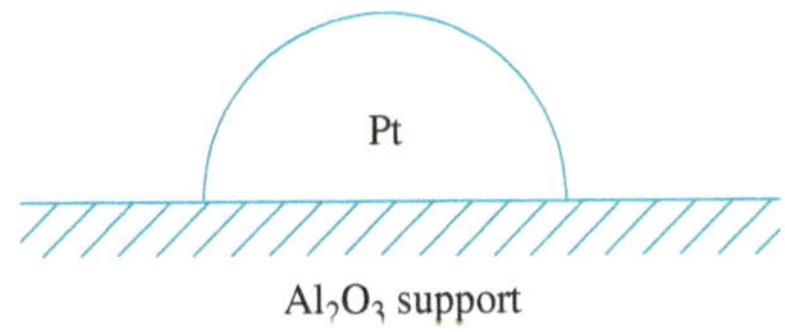

Figure 3.5:
Schematic representation of a supported catalyst.

Chapter 4 to the subjects of supports and catalyst preparation as well as catalyst characterization methods when we discuss in detail various methods of making and characterizing supported metallic catalysts. For the present purposes, all we need to say is that the metallic crystallites of a supported metal catalyst are generally located within the pores of the support and that these are large enough to allow the diffusion of the reactants to, and of the products away from, the metal surface. A typical supported metal catalyst, for example, Pt/alumina, can be depicted schematically as shown in Figure 3.5.

A well-known example of the use of supported Pt catalysts is in automobile exhaust catalysts.[12] In this case, the Pt is supported on a complex mixture of oxides including ZrO_2, CeO_2 and Al_2O_3, which in turn is supported on a honeycomb structure, generally consisting of a refractory oxide such as cordierite; alternatively, a metallic lattice made of a special alloy such as FeCrAlloy (containing Fe, Cr, Al and Y and fabricated into a honeycomb arrangement) is sometimes used. A very effective illustration of the operation of a three-way exhaust catalyst is given at http://www.dcl-inc.com/index.php?option=com_content&view=article&id=61&Itemid=72/

TASK 3.3 Car Exhaust Catalysis

Examine the literature describing the technical development of car exhaust catalysts. As we will later return (Chapter 8) to the subject of car exhaust catalysts to discuss the catalytic aspects of the various reactions occurring in the "catalytic muffler", your study should concentrate on the technical developments, from the initial introduction of the first emission control catalysts for conventional petrol and diesel engines to more recent developments in relation to catalysts for lean-burn engines. You should also examine alternative technologies such as the use of bio-fuels or hydrogen fuel cells. Starting points for your study could include sites of catalyst manufacturers such as http://ect.jmcatalysts.com/ and http://www.catalysts.basf.com/Main/mobile_emissions/ or of car manufacturers such as Ford, General Motors and Toyota. For an interesting early article on Johnson Matthey's work on exhaust purification catalysis, see G.J.K. Acres, B.J. Cooper, Platinum Metals Rev. 16 (1972) 74 (http://www.platinummetalsreview.com/dynamic/article/view/pmr-v16-i3-074-086/).

12 An equally important application of Pt catalysts is in hydrocarbon reforming. This will be discussed in Chapter 8

Many other examples of the importance of supported metal catalysts will be discussed in subsequent chapters, either in relation to methods of catalyst preparation or in relation to catalysts used in particular processes. It should be recognized that the support itself may not only be an inert material on which the metal is positioned: it may also contribute to the activity, selectivity and stability of the catalyst under operation conditions (see Box 3.6). As indicated above, the catalyst formulation may also include modifiers or promoters of various types. We will return to these topics later.

BOX 3.6 Evidence for the Participation of Oxygen of the Support

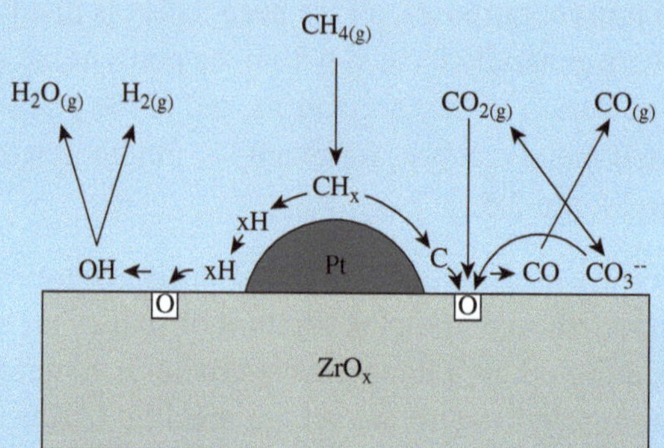

Experiments carried out using diffuse reflectance Fourier transform infrared spectroscopy (DRIFTS) have shown that the mechanism for the CO_2 reforming of methane over Pt/ZrO_2 catalysts probably takes place as shown in the scheme above. Carbon formed by the decomposition of methane on the Pt sites reacts with oxygen from the support in the proximity of the Pt crystallite and partially reduces the support. The oxygen is then replenished by reaction with CO_2 from the gas phase (A.M. O'Connor, F.C. Meunier, J.R.H. Ross, Stud. Surf. Sci. Catal. 119 (1998) 819)

3.5 Oxides

In just the same way that a metal is usually used under reducing conditions, oxides are generally used as active catalytic components either in oxidation reactions under oxidizing conditions or in reactions in which reduction of the oxide is prevented by kinetic or thermodynamic limitations.

In selective oxidation reactions, the catalytically active oxide frequently acts according to the so-called Mars–Van Krevelen mechanism (see Chapter 6, Section 5) in which an oxidation–reduction sequence occurs on the oxide surface. This generally requires that the metal ions of the oxide which constitutes the active phase has a variety of possible oxidation states. Examples of the oxides found in such catalysts include V_2O_5 (reducible to +4 and +3 states) and MoO_3 (reducible to +5 and +4 states). Mixed oxides, which are also commonly found, for example, BiMoOxides used for the selective oxidation of propylene (propene) to acrolein. We will return later in Chapter 6 to the subject of selective catalytic oxidation.

Non-reducible oxides generally function as inert supports that help to stabilize the metal particles (see above), examples being alumina (Al_2O_3), silica (SiO_2), magnesia (MgO) and zirconia (ZrO_2). However, even with these relatively stable oxides, there is some evidence that the interface between the metal and the support provides a special type of site and that the catalytic reactions occurring on the supported catalysts depend on sites on both the metal and the oxide surface near the interface, even utilizing oxygen ions of the support in the reaction. (See, for example, BOX 3.6 above)

Non-reducible oxides can also function as solid acids or bases. For example, alumina, especially when calcined at relatively low temperatures, has some acidic properties while MgO, CaO and BaO have basic properties. Silica-alumina, which can be produced by co-gelling the individual components, followed by washing and calcination, has relatively strong acidic properties and has been used as a hydrocarbon isomerization catalyst. In fluid catalytic cracking (FCC), the catalysts now used are zeolites (http://en.wikipedia.org/wiki/Cracking_(chemistry)/ http://en.wikipedia.org/wiki/Fluid_catalytic_cracking).

A recent application of alkaline earth oxides is in the trapping of SO_2 from automobile exhausts. For example, BaO is added to the catalyst formulation in the so-called Toyota system; the noble metal component of the catalyst, Pt, catalyses the oxidation of SO_2 to SO_3 and the BaO then adsorbs the SO_3 to give $BaSO_4$. A more detailed treatment of exhaust gas catalysis is given in Chapter 8.

BOX 3.7 Selective Reduction of NOx over V_2O_5/TiO_2 Catalysts

Selective catalytic reduction (SCR) of the NOx emissions[a] in the oxygen-containing stack gases from power stations using ammonia as reductant is generally carried out using V_2O_5/TiO_2 catalysts. We showed[b] that the reaction probably involves reduction of surface V5+ sites in steps (i) and (ii) and re-oxidation in steps (iii) and (iv):

$$V + O + NH_3 + NO \rightarrow V\text{-}OH + N_2 + H_2O \quad \text{(i)}$$

$$3V = O + 2NH_3 \rightarrow 3V\square + N_2 + 3H_2O \quad \text{(ii)}$$

$$V\square + 1/2\ O_2 \rightarrow V = O \quad \text{(iii)}$$

$$V\square + NO \rightarrow V = O + 1/2\ N_2 \quad \text{(iv)}$$

where V□ represents a surface vacancy or a *V*—OH group. A more detailed mechanistic description based on detailed isotopic exchange experiments has been reported elsewhere.[c] The SCR process is discussed in more detail in Chapter 8.

[a] NOx is a shorthand for a mixture of the oxides of nitrogen, predominantly NO. The whole area of SCR is summarized in a comprehensive review by H. Bosch, F.J.J.G. Janssen, Catal. Today 2 (1988) 369.
[b] H. Bosch, F.J.J.G. Janssen, F.M.G. van den Kerkhof, J. Oldenziel, J.G. van Ommen, J. R.H. Ross, Appl. Catal. 25 (1986) 239.
[c] F.J.J.G. Janssen, F.M.G. van den Kerkhopf, H. Bosch, J.R.H. Ross, J. Phys. Chem. 91 (1987), 5921, 6633.

3.6 Sulfides

The most common use of sulfides is in reactions in which a compound of sulfur is present, the reaction conditions favouring the formation of the sulfide from a metal or oxide present in the catalyst. The most common example is the $Co/Mo/Al_2O_3$ catalyst used for hydrodesulfurization (HDS):

$$RSH + H_2 \rightarrow RH + H_2S$$

The catalyst as prepared consists of a mixture of CoO and MoO_3 supported on a high area alumina such as γ-Al_2O_3. However, under the hydrodesulfurization conditions, the surface Co and Mo surface species transform to give a stable sulfide structure. Much has been written about the relative positions of the Co and Mo species but it is now generally accepted that the Co ions sit on the edges of MoS_2 platelets. We will return to a discussion of the process in Chapter 8.

TASK 3.4 Hydrodesulfurization Catalysts*

The company Haldor Topsøe A/S of Lyngby, Denmark (http://www.topsøe.com/ is at the forefront in the application of hydrodesulfurization catalysts as well as in other technologies for the chemical industry. Haldor Topsøe is also an increasingly rather unique example of a company which, in addition to marketing commercial catalysts and processes, also carries out important basic research underlying the processes which it markets. Study the Haldor Topsøe web site and follow up on a number of the processes and catalysts which Haldor Topsøe markets. Attempt to identify some of the companies competing with Haldor Topsøe in relation to these. Some of the important developments of Haldor Topsøe in the field of hydrotreating catalysis are based on the work of Henrik Topsøe, son of the founder of the company. Examine some of the publications of Henrik Topsøe and outline the science behind the "Brim" Technology (http://www.topsøe.com/research/BRIM_story.aspx)

* We will return to this subject in Chapter 8

3.7 Conclusions

We have seen in this chapter how a catalyst works and we have examined some aspects of the use of metals, oxides and sulfides as catalysts. We will now proceed to examine in more detail how a catalyst is made, characterized and tested.

CHAPTER 4

Catalyst Preparation

Chapter Outline

4.1 Importance of Active Surface Area and of Catalyst Structure

We have seen in Chapter 3 that a heterogeneous catalytic reaction proceeds on the surface of the catalytically active material by the making and breaking of bonds between the reacting species and the atoms of the catalyst surface. In discussing heterogeneous catalysis, we are therefore dealing essentially with *chemical processes occurring in two dimensions*, with the added complication that the reacting and product molecules have to

Heterogeneous Catalysis. DOI: 10.1016/B978-0-444-53363-0.10004-0

approach and leave the surface in a three-dimensional process. In order to optimize the rate of the reaction, the surface area of the catalyst must be high; if the surface is non-uniform in composition (e.g. with a supported metal, there are metallic sites as well as inert, or relatively inert, support sites), it must contain as many positions as possible at which the molecules involved in the process can be adsorbed. Further, as will be discussed in more detail in Chapter 5, all of the surface, external and internal, must be accessible to the reactant molecules approaching it from the gas phase (or from the liquid phase) and it must be possible for the products to get away from the surface. The situation for the reaction:

$$A + B \rightarrow P \tag{4.1}$$

discussed in the previous chapter is shown schematically in Figure 4.1. Each of the molecules A and B approaches the solid catalyst material from the gas phase, making a series of gas-phase collisions on the way. These collisions can be with other A molecules or with the species B, the product P or even an added inert gas. Once the molecule A reaches the solid, it then has to get to the active surface, which might be, for example, a metal crystallite embedded within a pore. Therefore, A has to diffuse down the pore and will make collisions with the walls of the pore as well as with gas-phase molecules; the relative number of surface collisions relative to gas-phase collisions will then depend on the pore diameter as well as on the reaction conditions (pressure and temperature). Once adsorbed on the active surface, the adsorbed species derived from molecule A can react with an equivalent species resulting from the diffusion and adsorption of molecule B to give adsorbed product P; P then desorbs and has to diffuse out of the pore and away from the surface. The rate of the reaction may still be

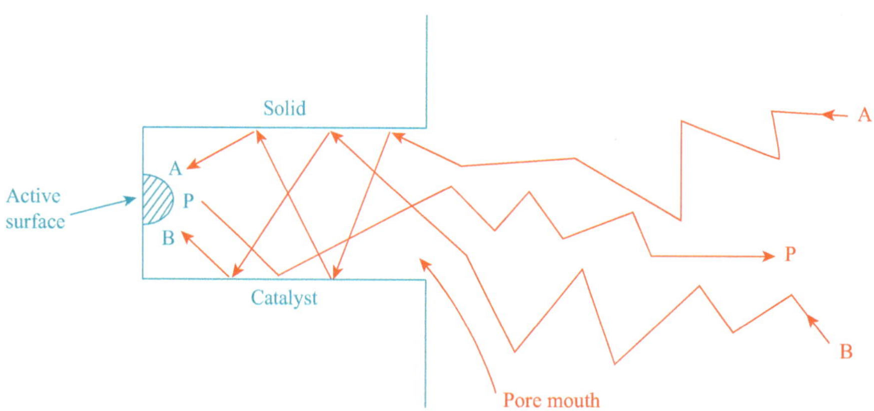

Figure 4.1:
Schematic representation of molecules A and B approaching and entering a catalyst pore, being adsorbed on the active surface within the pore, followed by the desorption of product P and diffusion of P out of the pore.

determined by the rate of adsorption of A or B, by the surface reaction of A with B or by desorption of product P; however, the rate may also be affected by the rates of diffusion to or from the surface and also into or out of the pore. Hence, in addition to having a high active catalytic surface, the catalyst must have easily accessible outer and inner surfaces; and, we shall see later that not all of the solid surface necessarily needs to be active.

The following sections will consider the creation of catalytic materials with high active surface areas in a form suitable for practical application and will also discuss the importance of the texture of the catalyst so formed. They will also discuss some of the more important methods of characterizing the catalysts prepared.

4.2 Catalyst Preparation

4.2.1 The Active Surface

An effective catalyst may have many different types of active surface species and each will be capable of catalysing either a single reaction or a whole range of related reactions. The most commonly encountered active species are metals, oxides and sulfides, as discussed in the previous chapter, but many other materials have also been shown to have catalytic activities. For example, as discussed previously (Chapter 3), metal species are most commonly used for reactions of hydrocarbons, examples being hydrocarbon reforming or steam reforming of natural gas. Oxides are used for different types of reaction, typically selective oxidation reactions; however, as the oxides can exhibit acidic or even basic properties, they are also used for reactions requiring either acids or bases, a typical example being catalytic isomerization of hydrocarbons. And sulfides can be used for reactions involving sulfur-containing molecules such a hydrodesulfurization. A word of caution is necessary here: all rules can be broken, especially in such a complex subject as catalysis, and there are many exceptions to these generalizations. We will now consider some of the most important methods of making and characterizing high surface area metals, oxides and sulfides and will consider how these materials are likely to be used in practice.

In order to obtain a high surface area and hence increase the rate at which a reaction occurs at the resultant surface, it is desirable to obtain a highly dispersed material (e.g. metal, oxide or sulfide). Figure 4.2 shows schematically what happens when a solid of cubic form and with side of length 1 cm is divided into smaller cubes of side 0.5 cm. The original cube has a surface area of 6 cm^2 while the eight smaller cubes have a total area of $8 \times 6 \times 0.5^2 = 12$ cm^2. A further subdivision gives 32 cubes of side 0.25 cm and a total area of 24 cm^2, etc. If the side of the cube is 1 nm, the area of material is now 6×10^7 cm^2 or 6000 m^2, an increase of area by a factor of 10^7. The same factor will apply to the increase the rate of a catalytic reaction occurring on the

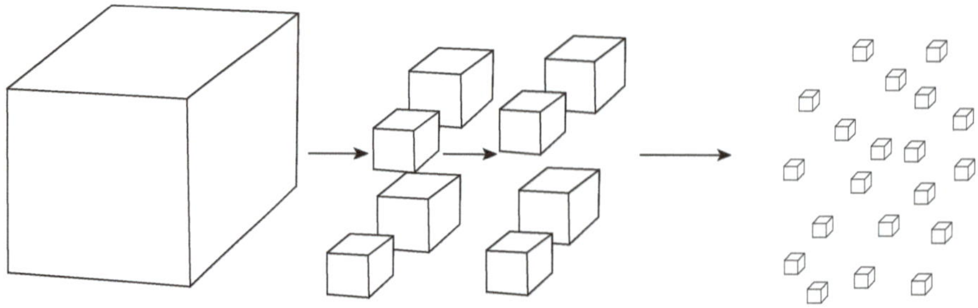

Figure 4.2:
Schematic representation of the increase of surface area caused by subdividing a cube of a solid material.

particles, at least as long as the reactant molecules have full access to the surface of each of the small (nano) particles.

TASK 4.1

Work out the area if the average dimension of the cubes formed by division is 5 nm (50 Å).

All materials have surface energies, these being equivalent to the surface tension of a liquid. Just as with a liquid, the surface area will tend to try to minimize itself in order to reduce the total surface energy. This means that the particles will have a tendency to coalesce to reform the original cube unless some way is found to stop that happening; this coalescence of the particles will be particularly important if the temperature is near the Tamman temperature for the solid (see Section 2.2 and also Section 4.3), the temperature at which atoms of that solid become mobile. The coalescence process is known as "sintering". While sintering is desirable in some applications, for example in ceramic manufacture, it must in the case of a heterogeneous catalyst be minimized since any sintering will cause loss of catalytic activity.

One way to prevent the sintering of a catalytically active component is to fix the particles of that component (which are not necessarily a single metal or oxide) onto another high-area surface ("support") in such a way that the particles cannot come into contact with one another

BOX 4.1 Minimisation of Surface Energy

A liquid will have a minimum surface energy if it exists as spherical drops as these will have the minimum surface area for any given drop size. Two separate drops will tend to coalesce to form a single one, hence minimizing the total surface energy, a good example being the coalescence of two drops of mercury sitting on a plane surface. Similarly, two soap bubbles brought into contact with one another will coalesce into a single bubble (see Task 4.2).

TASK 4.2

Two soap bubbles are connected to one another by a cylindrical tube.

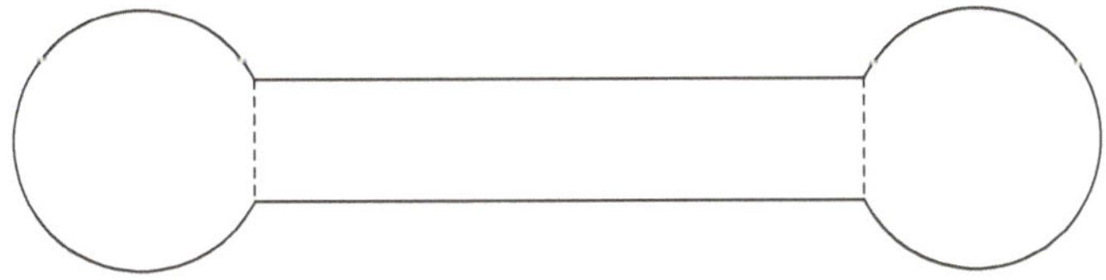

Will the bubbles remain the same size? If not, what will happen? (Note that the bubbles initially formed will each be approximately spherical, any deviation being due to the effect of gravity).

Answer: one will get smaller, and the other larger, until the surface areas of the two bubbles added together have reached a minimum. What will determine the diameter of the resultant single sphere?

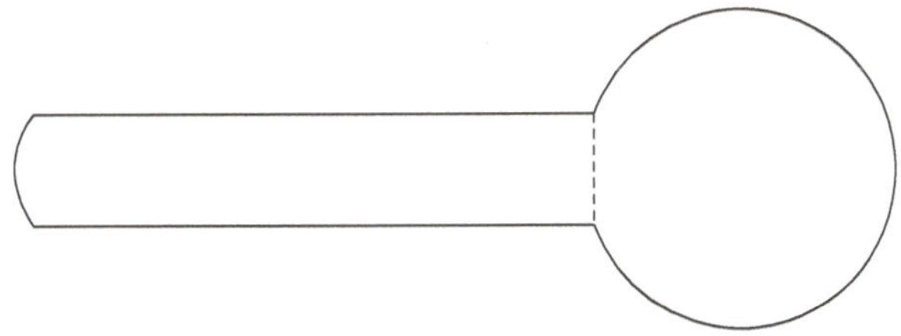

and that surface migration of the component atoms of the active particles across the support surface cannot occur. This is the principle of the "supported catalyst" in which the active component is situated on the surface of an inert and stable support material (see Chapter 3).

BOX 4.2 Support Effects

The support may or may not be considered to be totally inert. Frequently, there is some sort of interaction between the active component and the support; at one extreme, the active component is "glued" to what is otherwise an unreactive material, in the other extreme, there will be a "cooperation" between the active material and the support.

In another form of catalyst, the active component is the majority phase present and sintering is prevented by the presence of another minority phase that prevents the particles from coming into contact with one another and hence from coalescing. An example is the coprecipitated Ni-alumina catalyst with high Ni content that can be used for steam reforming or fat hardening; in this, alumina particles exist between the Ni crystallites (see Section 4.4.2).

We will now consider in turn the preparation of some typical supports and will then discuss the preparation (and, where necessary, pretreatment) of some typical supported and coprecipitated catalysts. We will also discuss some of the methods that can be used to characterize these catalysts.

4.3 Catalyst Supports

4.3.1 Introductory Remarks

No attempt will be made in this book to discuss the preparation and characterization of each and every type of known catalyst support. Instead, several typical supports will be considered and a number of generalizations, likely to be applicable to all types of supports, will be discussed.[1]

The most commonly used support materials are refractory oxides; these are metal oxides that have high melting points and are, therefore, very stable under most operating conditions. The constituent ions of an oxide become mobile at temperatures around the Tamman temperature (Table 4.1) that is approximately half the temperature of the melting point (in K). As shown in Table 4.1, the melting points of typical oxides used as supports range from about 1500 °C to about 2700 °C and so all these oxides will be usable up to temperatures of about 640 °C (silica–alumina) or higher.

Some oxide systems are complicated further by the possibility of phase transition under operating conditions. A number of oxides can be formed in several different phase modifications, some of which are stable over a wide range of temperatures, while others are not. (Further, a phase that may be stable under one set of operating conditions can change when subjected to other conditions, for instance, when exposed to water vapour.) As an example, TiO_2 can be made in three different forms: anatase, rutile and brookite. Anatase can have areas up to 80 $m^2\ g^{-1}$, while rutile, which forms at temperatures above about 550 °C, has a much lower area. The Tamman temperature for TiO_2 is around 800 °C (see BOX 4.3) and so phase transition in this case does not involve the bulk diffusion process associated with sintering as discussed above. A consequence is that each material used as a catalyst support must be examined in detail to make sure that it does not sinter or suffer from phase transitions under the conditions that will be encountered should that material be used in a particular catalytic reaction.

Another important parameter in determining which materials (oxides or other compounds) are used as supports is the texture of the solid. It is important that the support has a relatively high surface area and also that it has an accessible porous structure with reasonably large and

[1] For a more detailed discussion of catalyst preparation, the reader is referred to texts such as *Principles of Catalyst Development* by James T. Richardson (Plenum Press, New York and London, 1989), *Catalyst Preparation, Science and Engineering*, edited by John Regalbuto (CRC Press, Boca Raton, London, New York, 2007) and the proceedings of the series "International Symposia on the Scientific Bases for the Preparation of Heterogeneous Catalysts" held in Louvain la Neuve, Belgium and published by Elsevier Science Publishers in the Studies in Surface Science and Catalysis series, all of which are accessible through Science Direct and are indexed in Scopus.

Table 4.1: Values of the melting points and Tamman temperatures for a series of different oxide commonly used as catalyst supports. Several other oxides are also shown, which may in some cases be added as "promoters".

Oxide	Melting Point (K)	Tamman Temperature (K)	Oxide Type/Comments
γ-Al_2O_3	(2318)	(1159)	Acidic. Transfers into α-Al_2O_3 on heating (see text)
α-Al_2O_3	2318	1159	Amphoteric
SiO_2	1973	987	Acidic
SiO_2–Al_2O_3	1818	909	Very acidic
MgO	3073	1537	Basic
CaO	2853	1427	Basic
$CaSiO_4$	2407	1204	Basic
ZrO_2	2988	1494	Amphoteric
Cr_2O_3	2708	1354	Amphoteric
CeO_2	2873	1437	Amphoteric
La_2O_3	2588	1294	Amphoteric
$MgAl_2O_4$	2408	1204	Neutral
$ZnAl_2O_4$	2100	1050	Neutral

Note: The values shown in Table 4.1 are not precise and different sources give slightly different data.
Source: J.T. Richardson, Principles of Catalyst Development, Plenum, New York and London, 1989. The values agree reasonably well, but not entirely, with those give in the Handbook of Chemistry and Physics, *Chemical Rubber Company, 1974. However, the absolute values are not important in this discussion, only the orders of magnitude.*

regularly shaped pores. The surface area is of importance since the material to be supported also must have a high surface area to give it acceptable catalytic activities. The porous structure is also an important parameter since the molecules taking part in the catalytic reaction must be able to approach the active catalytic surface by diffusion through the pores and the products must also be able to diffuse out of the pores, as discussed in Section 4.1. (If the reaction only occurs on the outermost surface of the catalyst, then its "effectiveness factor" is well below unity and the consequence is that much of the active component in the catalyst is unused; if the active component is an expensive element such as Pt, this is highly undesirable.)

We will now examine some of the most important aspects related to the preparation and use of several of the most commonly encountered catalyst supports: alumina, silica, titania, zirconia and carbon.

4.3.2 Alumina, Al_2O_3

Alumina is one of the supports most commonly used in the preparation of heterogeneous catalysts. This is largely because it is relatively inexpensive and also because it can be prepared easily in forms that have high surface areas, coupled with readily accessible pore structures. However, it is also of importance because it is relatively inert and does not easily form compounds with the supported components (see further below). In certain applications, the fact

BOX 4.3 The Tamman temperature has also been used as an indication of the mobility of oxygen ions in selective oxidation catalysts. For example, based on the following data, Chen et al.* have argued that vanadia is more likely to have mobile lattice oxygen ions than are the molybdenum or tungsten oxides since it has the lowest Tamman temperature

Melting Point and Tamman Temperature of Various Metal Oxides

Metal Oxide	Melting Point (K)	Tamman Temperature (K)
V_2O_5	963	482
MoO_3	1068	534
Bi_2O_3	1098	549
CuO	1599	780
WO_3	1745	873
Fe_2O_3	1838	919
TiO_2	2128	1064
ZnO	2248	1124
NiO	2257	1129
Cr_2O_3	2538	1269

* Kaidong Chen, Alexis T. Bell, Enrique Iglesia, Kinetics and mechanism of oxidative dehydrogenation of propane on vanadium, molybdenum, and tungsten oxides, J. Phys. Chem. B 104 (2000) 1292–1299. http://iglesia.cchem.berkeley.edu/JPhysChemB_104_1292_2000.pdf

that the transition aluminas (see below) are acidic has some importance. It should be recognized that the preparation routes used for alumina as a catalyst support are also used in the preparation of the alumina that is smelted to give aluminium metal; as the aluminium industry is very large, there is a ready supply of alumina for other purposes.

The raw material used for alumina preparation is bauxite. This impure mineral is treated in the so-called Bayer Process in which the aluminium is dissolved in NaOH and separated from the impurities present in the original ore. The resultant sodium aluminate solution is then treated to precipitate alumium hydroxide (bayerite) from which alumina is subsequently made by calcination. (http://www.eaa.net/en/about-aluminium/production-process/) The alumina so prepared is relatively pure but may contain sodium ions as trace impurities. For use in catalysis, the alumina has to be particularly pure and is generally prepared from a solution of a pure salt such as the nitrate.[2] The salt solution is treated with an alkali[3],

[2] If another soluble salt such as the sulfate or chloride is used, the resultant alumina will contain trace sulfate or chloride ions and these may be detrimental to any catalyst made with such materials.

[3] If an alkali such as NaOH or Na_2CO_3 is used, the resultant alumina may contain traces of the associated cation and this is again undesirable.

e.g. ammonia, to give a precipitate of one of the hydroxides (gibbsite, bayerite or nordstrandite) or oxyhydroxides (boehmite or diaspore[4]) of aluminium, the actual structure produced depending on the conditions used for the precipitation step and the ageing of the precipitate.

The oxide or oxyhydroxide is then "calcined" to bring about dehydration and to form an alumina. Again, the conditions used for the drying and calcination steps can be crucial in determining the structure of the resultant alumina. In particular, the temperature of calcination has a very significant effect on the form achieved and its surface area. There have been many studies of the precipitation, drying and calcinations steps, one of the most definitive being that by B.C. Lippens (Thesis, T.U. Delft, 1961; B.C. Lippens, J.J. Steggerda, in: B.G. Linsen (Ed.), The Physical and Chemical Aspects of Adsorbents and Catalysts, Academic Press, 1970). The work of Lippens, which helped to clear up much of the confusion of nomenclature that had existed in the earlier literature, showed that the sequence for the production of the various transition aluminas and ultimately α-Al_2O_3 are as shown in the scheme of Figure 4.3.

As indicated above, which of the hydroxide or oxyhydroxide precursors is formed is very dependent on the preparation conditions. However, if the desired product is γ-Al_2O_3, boehmite, an oxyhydroxide, is first formed by hydrothermal treatment of one of the trihydrates and this is then calcined at about 723 K. Of the so-called transition alumina, γ-Al_2O_3 is the most commonly used as a catalyst support; the χ- and η- forms have also been used

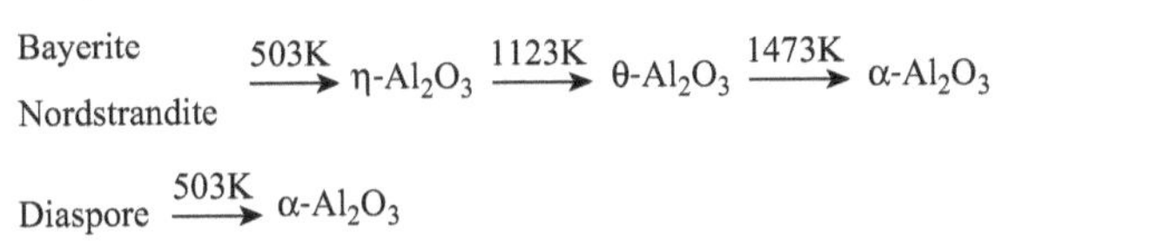

Figure 4.3:
The preparation of the transition aluminas. *Source: B.C. Lippens (Thesis, T.U. Delft, 1961; B.C. Lippens, J.J. Steggerda, in: B.G. Linsen (Ed.), The Physical and Chemical Aspects of Adsorbents and Catalysts, Academic Press, 1970.*

[4] Diaspore is also found as a naturally occurring mineral.

TASK 4.3

Examine the literature on aluminas using Scopus or the Web of Science. The thesis of Lippens is not cited in Scopus, nor is the book by Lippens and Steggerda. Hence, one needs to start by searching for other papers by B.C. Lippens (noting that his son, who published several papers later, had the same initials), particularly a series of papers in J. Catal. (3 (1964) 32–37, 38–43 and 44–49 and 4 (1965) 319–323.) Another useful lead is a more recent paper by C. Monterra, G.T. Magnacca, Catal. Today 27 (1996) 497–532. Papers by J.B. Peri on the use of spectroscopic methods to study oxide surfaces (see below) are also relevant.

as supports but the results obtained seldom differ too significantly from those obtained using γ-Al_2O_3, apart from the fact that these two forms appear to be stable up to rather higher temperatures.

All of the transition aluminas transfer into other modifications at temperatures higher than about 873 K and all are transformed into α-Al_2O_3 (corundum) at temperatures around 1473 K; the relative rates of transition depend markedly on the degree of crystallinity of the precursors used in the preparation. Corundum has a well-crystallized spinel structure while the transition aluminas are described as having deformed pseudo-spinel structures. Lippens and other authors are in agreement that the structures of the transition aluminas differ in the relative occupation of the octahedral and tetrahedral sites.[5]

The transition aluminas all have relatively high surface areas, typically several hundred $m^2\ g^{-1}$, and are used for the preparation of catalysts, which, in general, are not used at temperatures much above 500 °C. They all have acidic surfaces characterized by the presence on their surfaces of significant concentrations of hydroxyl groups formed either during the decomposition of the precursor or by interaction subsequently with water from the atmosphere:

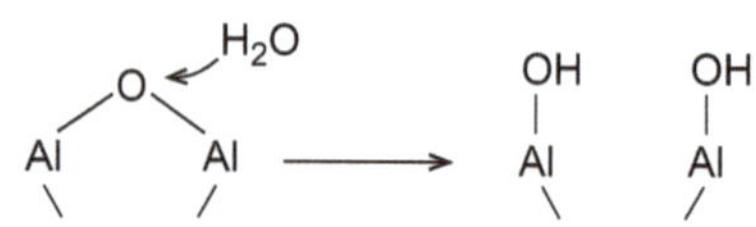

(4.2)

The surface AlOH species so formed have Brönsted acidic character. In other words, they give rise to protons on dissociation and hence act as solid acids. The dehydrated surface shown as the precursor of the reaction shown in Eq. (4.2) is a Lewis acid. If the surface is not fully hydrated, or becomes partially dehydrated on calcination, the surface has a mixture of Lewis

[5] A recent comprehensive study of the structure and properties of γ-Al_2O_3 has been carried out by G. Paglia. The work is described in detail in his Ph.D. thesis available on-line at http://adt.curtin.edu.au/theses/available/adt-WCU20040621.123301/unrestricted/01front.pdf. The thesis also contains a list of the papers based on the work.

and Brönsted acid character. It is this acidity that makes alumina particularly suitable as a support for the preparation of heterogeneous catalysts: in addition to providing "sites" at which precursors of catalytic components may be adsorbed, as described later, these OH groups have acidic character, which may contribute to the catalytic behaviour of the final catalyst. An important paper on the behaviour of alumina surfaces upon dehydration is by J.B. Peri who studied the dehydration of γ-Al_2O_3 using infrared and gravimetric methods.[6] He showed that OH species persist on specific sites on the surface even after dehydration at high temperature.

The forms of alumina other than γ-, χ- and η-Al_2O_3 are seldom used as catalyst supports as their surface areas are generally much lower than those of the transition aluminas. (There is of course the chance that some of these crystal structures may be formed in a catalyst which has been exposed to operation at higher temperatures, this usually resulting in deactivation; see Section 4.4.2 below.) Corundum is a very hard, refractory material with a very low surface area that generally finds application as a ceramic. However, porosity can be built into the structure and such materials have been used as matrices in some catalyst formulations (see Section 4.4.2.3 below).

Before proceeding to describe the preparation of catalysts using high-area supports, we will first discuss some other support materials used in catalyst formulations.

4.3.3 Silica, SiO_2

Silica has many applications, these including use in glass and whitewear ceramics, as a desiccant, as a component in toothpaste and scouring powders and similar preparations, and as a component of certain types of food (e.g. as a thickener in soups), as well as being used as a catalyst support (http://en.wikipedia.org/wiki/Silicon_dioxide). The silicas most used as catalyst supports are generally the same as those used as desiccants as they have high surface area and significant pore volumes.[7]

Typical desiccant silica is made by dissolving silica sand (quartz) in hot caustic soda:

$$SiO_2 + 2NaOH \rightarrow Na_2SiO_3 + H_2O \tag{4.3}$$

The sodium silicate is then reacted with acid (usually sulfuric acid) to form silicic acid and sodium sulfate:

$$Na_2SiO_3 + H_2SO_4 \rightarrow H_2SiO_3 + Na_2SO_4 \tag{4.4}$$

[6] J.B. Peri, J. Phys. Chem. 69 (1965) 211, infrared and gravimetric study of the surface hydration of γ-Al_2O_3

[7] Fumed silicas, formed by burning $SiCl_4$ in an air/hydrogen flame: $SiCl_4 + 2H_2 + O_2 \rightarrow SiO_2 + 4\,HCl$ have also been used.

The silicic acid formed then polymerizes to form long silicate chains by a polymeric chain growth mechanism involving further condensation steps:

$$\mathrm{H_2SiO_3 + H_2SiO_3 \rightarrow HO\text{-}\underset{\underset{O}{\|}}{\overset{\overset{O}{\|}}{Si}}\text{-}O\text{-}\underset{\underset{O}{\|}}{\overset{\overset{O}{\|}}{Si}}\text{-}OH + H_2O \rightarrow HO\{\underset{\underset{O}{\|}}{\overset{\overset{O}{\|}}{Si}}\text{-}O\}_nH} \tag{4.5}$$

The chains interact with one another forming first sols that then solidify to give hard gels.[8] The higher the concentration of the silica sols, the harder the resultant gels. Particles of these hard gels are those used in the many domestic cleaning preparations mentioned above.

The times taken for polymerization and subsequent gel formation ("gelation") depend strongly on the conditions: pH, concentration and temperature. In the industrial process, in which highly concentrated solutions are used, the mixing of the acid and the sodium silicate have to be carried out extremely rapidly in order to avoid conditions under which the gelation process occurs simultaneously in the mixer, a potential disaster since the gel is extremely hard and unreactive. This can be understood by reference to Figure 4.4, which shows schematically the gelation time (t_g, on a logarithmic scale) versus the pH of the gel.[9] It can be seen that if the pH of the gel is about 7, the gelation time is less than a minute. However, if the pH is closer to 1, the gelation time may be many hours. The pH of the sodium silicate used for the preparation

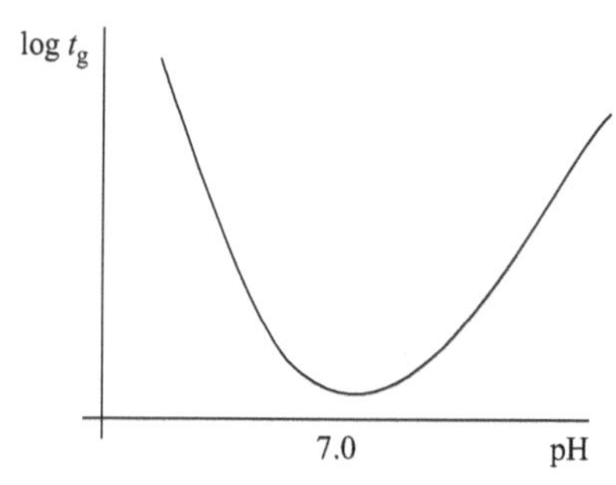

Figure 4.4:
Effect of pH on gelation time for a silica sol.

[8] This polymerization process of silicic acid was once used in a domestic product: the so-called "water-glass" used for preserving hens eggs. The silicic acid was provided as a powder and was made up as a "solution" (more correctly, a sol) in water. The eggs were then immersed in the sol and this gradually gelled, providing a matrix through which oxygen from the air could not penetrate, preserving the freshness of the eggs. As the concentrations were quite low, the gel was never hard enough to prevent eggs being removed easily when required for use. Inserting one's hand was sufficient to break the local gel structure; this reformed once one's hand was removed.

[9] R.K. Iler (1979). The Chemistry of Silica. Plenum Press. ISBN 047102404X. See also http://en.wikipedia.org/wiki/Silicon_dioxide; this gives a useful description of the uses of silica as well as information on structures, etc.

will be high (the exact value depending on the concentration of alkali added to dissolve the silica sand, Eq. 4.3). If only enough acid is added to neutralize the sodium silicate (Eq. 4.4), the pH will be close to the minimum (Figure 4.4) and gelation will occur almost instantaneously; if the concentrations are high, the resultant gel will be rock-hard and impossible to remove. To prevent this happening, a significant excess of acid is added (with rapid mixing to ensure that the pH throughout is uniform) in order that that the resultant pH is 4.0 or less. The gelation time is now sufficiently long to allow the sol to be removed from the mixing vessel before it can gel.

The gel formed is a mixture of a solid matrix of interlinking polymeric silica chains, water and sodium sulfate. The sodium sulfate (or other salt if another acid is used) is removed by washing and the gel is then dried to remove excess water. This procedure sounds simple but the properties of the resultant anhydrous gel are very dependent on the final pH of the gelation process as well as on the pH, the temperature and the time of any ageing process that follows; they also depend strongly on the temperature of the dehydration step (drying and calcination) applied after washing the material. If the ageing is carried out at relatively high pH, the surface area of the resultant gel can be as low as 200 $m^2\ g^{-1}$ but if the pH is low, the area can be as high as 900 $m^2\ g^{-1}$. The pore volumes and diameters are also strongly dependent on the ageing, washing and calcinations conditions. As a result, a large range of different materials is available commercially. It is interesting to note that a variety of different physical forms are also available and that some of these can be prepared making use of a well-controlled gelling process; for example, uniform spheres can be obtained if spherical drops of the sol are extruded into an inert matrix (e.g. an immiscible oil) in which the polymerization process can be completed prior to washing, drying and calcination as before.

While the pore structure of the gel is of importance in determining the properties of the catalysts made using the gel, the dehydration/calcination step is of importance in determining the way in which the gel interacts with metal species during the preparation of the catalyst. J.B. Peri, using infrared spectroscopy and other methods, studied the behaviour of the hydroxyl species on the surface of a silica gel as a function of dehydration temperature.[10] He showed that pairs of OH groups are attached to each surface silicon atom even after drying at 600 °C. Dehydroxylation at very high temperatures results in the creation of isolated OH species. Although the surface hydroxyl groups on silica are reasonably acidic, a more acidic solid is obtained by co-gelling silica and alumina. The resultant material, often shown as $SiO_2{-}Al_2O_3$, is used alone as a cracking catalyst or together with other components when an acidic support is needed.[11] Solids with zeolitic structures are now more commonly used for these purposes (see Section 4.3.4).

---

[10] J.B. Peri, Infrared study of OH and NH_2 groups on the surface of a dry silica aerogel, J. Phys. Chem. 70 (1966) 2937; see also J.B. Peri, A.L. Hensley, The surface structure of silica gel, J. Phys. Chem. 72 (1968) 2926.

[11] J.B. Peri, Infrared study of adsorption of carbon dioxide, hydrogen chloride and other molecules on acid sites on dry silica-alumina and gamma-alumina, J. Phys. Chem. 70 (1966) 3168.

TASK 4.4

Look up J.B. Peri in Scopus or the Web of Science and follow up his papers that relate to the use of spectroscopic methods in determining the surface structures of alumina and silica. Following up on the papers on alumina will lead to a paper by C. Morterra, G. Magnacca, Catal. Today 27 (1996) 497; you should include this in your further searches. This in turn will lead to a monograph on IR spectroscopy in catalysis by J. Ryczkowski, Catal. Today 68 (4) (2001). Document your search.

In concluding, it is worth noting two factors that are of some importance in relation to silica as a catalyst support. On the positive side, it should be recognized that silica exhibits no phase transitions on heating and hence one is only limited by the possibility of sintering; with a Tamman temperature of 987 K (Table 4.1), one can expect to be able to use silica as a support for reactions occurring up to about 700 °C. However, on the negative side, one has to recognize that silica is not an inert solid. For example, it can react with water vapour to form a volatile hydrosilicate at higher temperatures and so it cannot be used in, or even as a component of, catalysts to be used in steam reforming or related processes. It can also react with alkali or alkaline earth metals to form the corresponding silicates and so it cannot be used in any situation where such species might be present since the silicates formed are often mobile and can cause, inter alia, severe pore blockage.

4.3.4 SiO_2—Al_2O_3 and Zeolites

As discussed in the previous section, SiO_2—Al_2O_3 is an amorphous mixed oxide material prepared by mixing silica and alumina sols. According to Richardson, the former can be a 30% mixture of water glass and water (see preparation of silica above) and the latter a mixture of 4 M HCl and 0.5 M $Al_2(SO_4)_3$. These are rapidly mixed in the ratio 2:1 at a temperature of 5 °C and rapid gelation occurs. After ageing the gel, it is treated with 2% $Al_2(SO_4)_3$ solution to ion exchange the Na^+ ions trapped in the gel and then washed free of SO_4^{2-} ions; it is then dried and calcined at 550 °C. It is a much more acidic material than either of its components due to weakening of surface OH bonds by the proximity of the Si^{4+} and Al^{3+} centres:

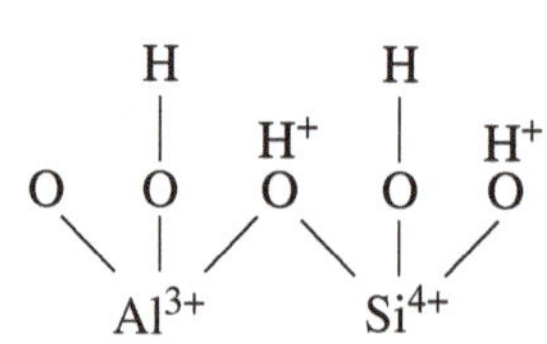

In this structure, the Si^{4+} ion replaces the tetrahedral Al^{3+} with a more electropositive centre, thereby weakening the OH bond and increasing the acidity. These silica-alumina materials have been used as cracking catalysts or as acidic supports.

Zeolite materials are crystalline modifications of silica-alumina. They are prepared by hydrothermal treatment of co-gels of silica and alumina in an autoclave at up to 10 atm. and at temperatures between 90 °C and 180 °C, in the presence of OH^- ions, a suitable organic template (e.g. an organic amine or an alkyl ammonium compound) as well an appropriate seed crystals. A wide variety of different structures have been prepared and characterized, these having a large range of Si/Al ratios as well as differing pore geometries and sizes.[12] Alumino-phosphate molecular sieves may also be synthesized and other oxides such as TiO_2 can also be incorporated. Zeolites are now used in many different applications, many of which are discussed in chapters by different authors in the book referred to in footnote 12.

TASK 4.5

Van Bekkum and his colleagues (J.C. Jansen et al., Microporous and Mesoporous Materials, 21 (1998) 213) have reviewed the preparation and potential uses of zeolitic coatings in catalysis. Read this review and examine some of the papers that have quoted it, paying particular attention to those that describe the preparation of zeolitic membranes.

4.3.5 Titania, TiO_2

Titania is used as a support in certain reactions, such as in the selective catalytic reduction (SCR) of NOx (see Box 3.6 and Task 4.7; see also Section 8.5.1) for which it is used as a support for vanadia. However, because of its semiconductor properties, it can also be used on its own in photocatalytic processes.[13] The most common use of TiO_2 is as a pigment for paints and it is made by one of two processes: hydrolysis of $TiOSO_4$ ("Sulfate Route") or by the vapour phase oxidation of $TiCl_4$ ("Chloride Route"). The sulfate process is no longer favoured because of the problems of disposing of the large quantities of impure concentrated sulfuric acid formed as a by-product.

As discussed above, TiO_2 may exist in several structural forms: rutile, anatase and brookite. The anatase form is more usually used as a support and the surface area of typical materials is of the order of 80 $m^2\ g^{-1}$, relatively low compared with alumina and silica. It transforms to rutile, losing surface area, at about 550 °C. This means that the applications of titania as

[12] There are many possible sources of additional information on this topic, for example, H. van Bekkum, E.M. Flanigen, P.A. Jacobs, J.C. Jansen (eds.), Introduction to Zeolite Science and Practice, Stud. Surf. Sci. Catal. 137 (2001).

[13] See, for example, S. Malato, P. Fernandez-Ibanez, M.I. Maldonado, J. Blanco, W. Gernjak, Decontamination and disinfection of water by solar photocatalysis: recent overview and trends, Catal. Today 147 (2009), 1–59.

a support are limited to reactions operating at relatively low temperature such the SCR reaction mentioned above. Much of the titania used as a catalyst support (and also for studies of photocatalysis by the oxide itself) in work reported in the literature has been based on a pigment grade titania, given the name P25, supplied by Degussa, now Evonic Industries http://corporate.evonik.com/en/Pages/default.aspx. However, many other suppliers of titania exist.

4.3.6 Zirconia, ZrO_2

Zirconia has recently attracted much attention as a catalyst support, partly because it is more chemically inert than the more traditional supports such as silica or alumina and also partly because it has been shown to have specific interactions with the active phase, giving rise to differences in the activities and selectivities of the resultant catalysts. Zirconia may be formed with a number of different phase structures. The unstable tetragonal phase is formed by gel precipitation but the resultant material is unstable on calcination at higher temperatures, the area dropping by as much as 97% on heating to 850 °C. The sintering was found to occur by a surface transport mechanism resulting in a phase transformation to the monoclinic form, the latter being the most stable phase at normal temperatures of usage.[14] Mercera et al. showed that it is possible to prepare pure samples of the monoclinic phase by careful gel precipitation followed by careful washing in ethanol.[15] The monoclinic phase so formed was stable up to temperatures of 1170 °C. However, there was still a significant reduction of surface area (from about 72 to 17 $m^2\ g^{-1}$) when the calcination temperature was increased from 450 °C to 900 °C (see Figure 4.5). The stability of the monoclinic phase could be improved significantly by the addition of species such as Ca, Y or La, as shown in Figure 4.6; as little as 20–50% of a theoretical monolayer of one of these ions was sufficient to prevent sintering by preventing significant surface mobility of the Zr ions. Figure 4.6 shows that the area of the zirconia, particularly with the addition of the rare earth oxides of Y and La, was now relatively unaffected as long as the calcination temperature was below about 700 °C. Figure 4.7 shows the effect of the concentration of the most effective additive, La, as a function of the La_2O_3 concentration for a series of different calcination temperatures; up to a calcination temperature of 700 °C, as little as 3% La had a very significant stabilizing effect; even after calcination at 900 °C, the addition of La caused a significant improvement in surface area (with values up to ca. 35 $m^2\ g^{-1}$ compared with about 17 $m^2\ g^{-1}$ for the unpromoted material). We will return later to discuss the use of ZrO_2-supported catalysts in several different applications.

[14] P.D.L. Mercera, J.G. van Ommen, E.B.M. Doesburg, A.J. Burggraaf, J.R.H. Ross, Zirconia as a support for catalysts: Evolution of the texture and structure on calcination in air, Appl. Catal. 57 (1990) 127–148.

[15] See, for example, P.D.L. Mercera, J.G. van Ommen, E.B.M. Doesburg, A.J. Burggraaf, J.R.H. Ross, Zirconia as a support for catalysts. Influence of additives on the thermal stability of the porous texture of monoclinic zirconia. Appl. Catal. 71 (1991) 363–391.

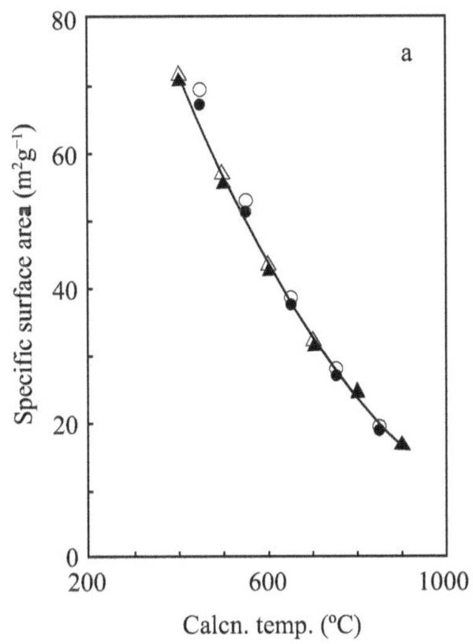

Figure 4.5:
The effect of calcination temperature on the specific surface area of an un-doped monoclinic zirconia sample. *Source: Mercera et al., Appl. Catal., 71 (1991) 363–391. Reproduced with kind permission of Elsevier.*

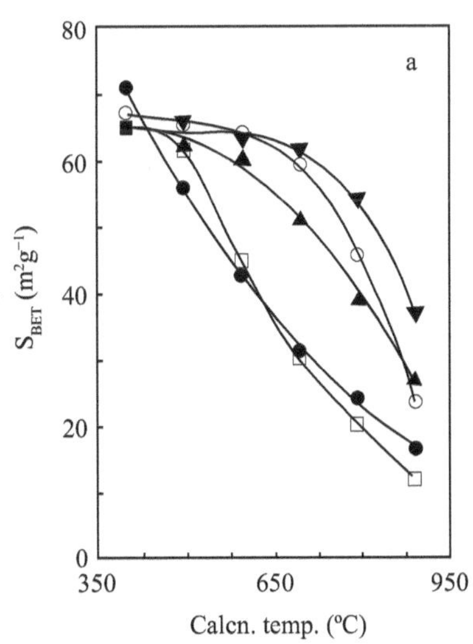

Figure 4.6:
The influence of various additives (ca. 5 mol%) on the thermal stability of monoclinic ZrO_2. ●, undoped; □, MgO; ▲, CaO; ○, Y_2O_3; ▼, La_2O_3. *Source: Mercera et al., Appl. Catal., 71 (1991) 363–391. Reproduced with kind permission of Elsevier.*

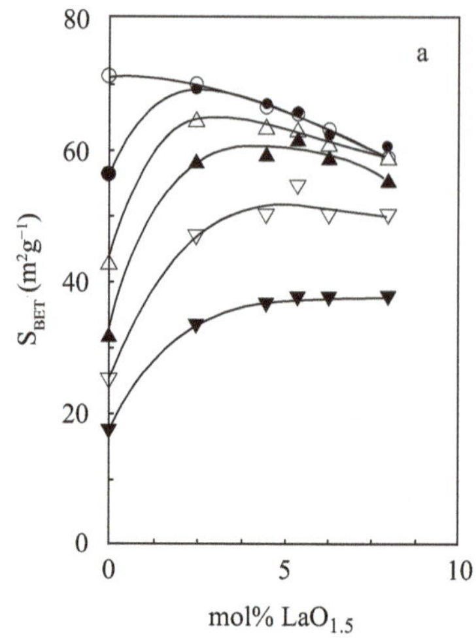

Figure 4.7:
The effect of the La_2O_3 concentration on the surface area of ZrO_2 as a function of calcination temperature: ○, 400; ●, 500; Δ, 600; ▲, 700; ∇, 800; ▼, 900°C. *Source: Mercera et al., Appl. Catal., 71 (1991) 363–391. Reproduced with kind permission of Elsevier.*

TASK 4.5

Examine the literature on the preparation, characterization and uses of zirconia. A useful starting point is to examine the work of P.D.L. Mercera and his colleagues and the references that cite their papers.

4.3.7 Carbon

Carbon is often used as a catalyst support, particularly in very specific organic reactions, in the so-called Fine Chemical Industry (see Chapter 8). The resultant catalysts are frequently used in hydrogenation reactions that are carried out in the liquid phase and it would appear that carbon-supported materials are particularly suited to such conditions, perhaps because they are less likely than many other supports to interact with the liquids used. There is also some indication that the pore structure is such that physical adsorption of the reactant molecules occurs in the narrow pores, bringing the reactant into closer proximity to the active site (often metal micro-crystallites). The "activated" carbons used as supports are most commonly made from naturally occurring materials such as wood, coal and peat.[16] A consequence of the preparation routes used (normally pyrolysis), the resultant material can have surfaces

[16] One of the more commonly used active carbon supports is made by the company Norrit in the Netherlands (http://www.norit-ac.com) using peat from the peat bogs in that country. Norrit carbon was originally sold predominantly to remove the colour from sugar syrup in the production of white sugar.

containing a wide variety of different surface entities (carboxyl, keto, hydroxyl, etc.), the concentrations of each depending on the preparation and pretreatment conditions.

4.3.8 "Forming" of the Support

Before the introduction of the active phase into the chosen support, it is generally "formed" by either extrusion or pelletting to give a suitably strong matrix in which to place the catalytically active material. In the case of extrusion, a suitable paste of the support material is made and this is extruded through appropriately shaped dyes before being chopped into suitable lengths for the application envisaged. The extrudates are then dried and "fired" or "calcined" at a temperature high enough to impart the desired strength.[17] Pelletted supports are generally made from dried powders to which suitable lubricants (e.g. graphite) have been added prior to compression in the desired shape using a tabletting machine, this again being followed by drying (if necessary) and calcination. This topic will receive some further attention in Chapter 7.

We will now proceed to discuss the methods commonly used to prepare supported heterogeneous catalysts by impregnation or deposition/precipitation.[18] An alternative method of preparation, coprecipitation, will be described in the subsequent section.

4.4 Supported Catalysts

The active catalytic phase of a catalyst may be introduced into the support structure by either impregnation or by deposition-precipitation prior to drying and calcination. In both cases, a suitable solution is made up and introduced into the pores of the support. As noted above, the support is generally preformed to the desired shape.

4.4.1 Impregnated Catalysts

In the case of impregnation, the method depends on an interaction occurring between the surface of the support and the species in solution. This interaction may be as a result of ion exchange with the acidic hydrogen ions of surface hydroxyl groups or it may be more of a physical nature, especially if the concentration of impregnation species exceeds the

[17] S.P.S Andrew once described a typical industrial test for a mechanically strong catalyst: "take it into the Gents' washroom and throw it into the air; if it lands on the tiled floor without breaking into bits, it is strong enough".

[18] Catalyst preparation used to be described as an art rather than a science and it is only within the last 40 years that a more scientific approach to catalyst preparation has become more prominent. This is partly because of the initiation of the series of International Symposia on the Scientific Bases for the Preparation of Heterogeneous Catalysts, which commenced in 1975 (Volume 1 of Studies in Surface Science and Catalysis) and also because of the realization that someone studying a particular catalyst or catalytic system cannot simply take it off the shelf but needs to have a full characterization of the specific material used.

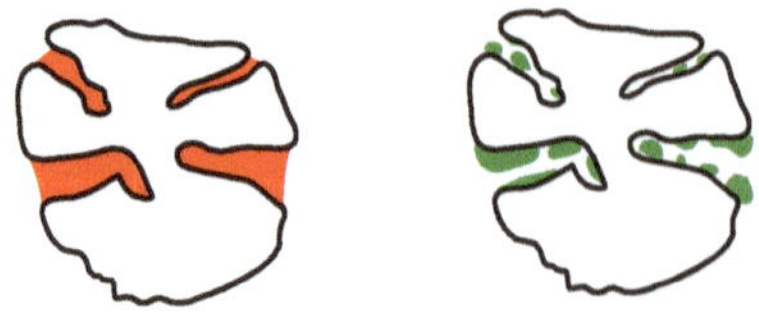

(a) (b)

Figure. 4.8:
Pore filling during wet impregnation and the effect of subsequent drying.

concentration of surface hydroxyl groups. If the volume of impregnation solution introduced is equal to or less than the volume of the pores, the technique is known as "pore volume" or "dry" impregnation and the method depends more on retaining the species within the pores during drying rather than on a specific interaction; however, if the volume of solution is in excess of that required to fill the pores, the technique is known as "wet impregnation" and specific interactions play a much more important role. The advantage of dry impregnation is that the weight of added component incorporated in the catalyst can easily be controlled. However, the resultant material may not be as uniform as that prepared by wet impregnation; for example, there may be regions of the catalyst with larger pores in which there will be larger concentrations of catalytically active material than in others. As is shown schematically in Figure 4.8, the larger pores will take up larger volumes of solution and the result is that there is likely to be larger crystallites of the precursor in these pores. In addition, if the drying process is not carried out carefully, the catalytically active material may become concentrated at the mouth of the pores or may even move onto the outer surface of the support (see below). If the support has been preformed, the distribution of pores and hence the distribution of the active material in the calcined catalyst will also be dependent on the parameters of the pelletting process.[19]

Wet impregnation is more difficult to control than dry impregnation. If the concentration of solute ions is low, then all of the solute ions may be taken up at the mouth of the pores (e.g. by ion exchange with acidic surface groups on the support surface near the pore mouths) and none will be found in the bulk of the support. The formation of an "egg-shell catalyst" with a high concentration of the active component at the outer surface of the pellet may of course be desirable either if this component is expensive (e.g. Pt or Rh) or if the reaction occurs predominantly near the outer surface of the catalyst pellets as a result of pore diffusion limitations. If, however, the reaction is totally kinetically controlled and all of the bulk of the catalyst is utilized (the "Effectiveness Factor" is 1.0), it is desirable that the catalytically active component is spread throughout the pellet. In that case, multiple impregnations may be

[19] If the impregnation step is carried out on the powdered support prior to pelletting, more even distributions will be obtained as long as the particle size is quite uniform. However, changes occurring during the subsequent pelletting process will then be difficult to control.

necessary to ensure that the active component is present in all parts of the support and in sufficiently high concentrations.[20]

TASK 4.6

In the above discussion of the preparation of impregnated catalysts, many aspects of the process have been omitted. For example, we have not considered the importance of the chemical step occurring when impregnation is carried out. Interaction with surface acidic groups can play a role (Brønsted vs. Lewis sites) and the structure of the double layer forming at the support-solution interface can also determine the ease with which reaction occurs. Paying attention to such parameters, follow some of the literature on the preparation of catalysts. For example, using the name K.P. de Jong (University of Utrecht), use Scopus or Web of Science to identify some relevant papers to study. You could start with the paper from his group on the preparation of supported Pd catalysts (M.L. Toebes et al., J. Molec. Catal. 173 (1997) 281) and then read some of the literature quoted there or quoting this work. Other authors whose work could be included in your study are M. Che (see, e.g. Chem. Rev. 97 (1997) 305), C. Parego and P. Villa (Catal. Today 34 (1997) 281) or S.P.S. Andrew (Chem. Eng. Sci. 36 (1981) 1431). The paper by Parego and Villa, while being a general description of catalyst preparation, also gives details of industrial practice; the paper by Andrew concentrates on examples of the preparation of industrial catalysts.

TASK 4.7

A particularly well-studied system is the case of vanadia supported on titania for which various different preparation methods have been used. For example, the catalysts examined in a paper by Bond et al. (Brunel University) and Gellings et al. (University of Twente) included aqueous impregnation of $(NH_4)_2[VO(C_2O_4)_2]$ and reaction with VOCl3, $VO(OR)_3$ (R=iBu) and $VO(acac)_2$ in organic solvents (acac = acetonylacetonate (Appl. Catal. 22 (1986) 361). Using this reference as a starting point, examine the literature on vanadia/titania catalysts, paying particular attention to the effect of preparation method on the catalyst structure. Your study should include a more recent paper by I.E. Wachs and B.M. Weckhuysen (Appl. Catal. A 157 (1997) 67) on the structure and reactivity of vanadium oxide species on oxide supports.

With both dry and wet impregnation, the drying process carried out after the impregnation step is very important. If the drying process is carried out too quickly, the liquid contained in the pores may be driven out of them by vapour forming deep in the pores, this resulting

[20] It is possible to make catalysts in which the distribution of the active component is at a maximum somewhere between the outer surface and the interior or even that the maximum concentration is found right at the centre of the pellet. Such distributions may have some practical advantages. (S.Y. Lee, R. Aris, The distribution of active ingredients in supported catalysts prepared by impregnation, Catal. Rev. Sci. Eng. 27 (1981) 207.)

in deposition of the active phase on the exterior surface of the support. This is undesirable since such material is unlikely to adhere sufficiently strongly to the support and so "dusting" may occur in use: accumulation of a dust of the active component down-stream of the catalyst bed, this sometimes even resulting in blockage of the exit of the reactor. If the catalyst is to be calcined before use, the way in which this is carried out is also important. If the calcination step involves the decomposition of a salt such as a nitrate, the nitrogen oxides formed should be flushed out of the catalyst by a carrier gas to prevent the occurrence of undesired reactions. Even the elimination of any water vapour formed during the decomposition step should be carefully controlled as water vapour can encourage sintering in some oxides. Finally, if the catalyst has to be reduced before use, the reduction is best carried out slowly, often with a gradually increased temperature of reduction.[21]

4.4.2 Coprecipitated Catalysts

4.4.2.1 Precipitation of single cations

Before considering coprecipitation, we will discuss briefly the precipitation of individual ions. The solubility of a compound, A^+B^-, is a function only of the temperature and is determined by the solubility product, K_s, a thermodynamic quantity:

$$K_s = a_A^+ \cdot a_B^-$$

where $a_A{}^+$ and $a_B{}^-$ are the solubilities of the cation and anion respectively. (For our purposes, B^- is often a hydroxide ion, OH^-). For hydroxide compounds containing multicharged cations, the expressions for the solubility product are more complex; as an example, the solubility product for aluminium hydroxide ($Al(OH)_3$), is given by

$$K_s = a_{Al}^{3+} \cdot (a_{OH}^-)^3$$

Table 4.2 shows the values of K_s for a number of metal hydroxides commonly encountered in catalyst preparation. All of the hydroxides are highly insoluble and so we can conclude that if OH^- ions are added to a solution containing one of the metal ions (cations) shown, there will be effectively 100% precipitation of the corresponding hydroxide.

[21] Reduction is most frequently carried out in the reactor where the catalyst is to be used as there are then no problems of re-oxidation prior to use. However, there are cases in which a pre-reduced catalyst is supplied by the catalyst manufacturer. In such a case, some method has to be worked out to passivate the catalyst so that it is safe to transport and so that it can readily be reactivated in the reactor of the purchaser. Passivation is often carried out by very careful re-oxidation of only the surface layer or layers by exposure of the catalyst to a low concentration of oxygen. (Often, the contamination level of oxygen in very pure nitrogen is sufficient for this purpose.) It should be noted that a pre-reduced catalyst is a potentially dangerous chemical: it can burn "pyrophorically" if the rate of re-oxidation is sufficiently high and uncontrolled.

Table 4.2: The solubility product (*Ksp*) for some common hydroxide compounds at 298 K.

Compound	Formula	Ksp (298 K)
Aluminium hydroxide	$Al(OH)_3$	3×10^{-34}
Cobalt hydroxide	$Co(OH)_2$	5.92×10^{-15}
Copper (II) hydroxide	$Cu(OH)_2$	4.8×10^{-20}
Iron (II) hydroxide	$Fe(OH)_2$	4.87×10^{-17}
Iron (III) hydroxide	$Fe(OH)_3$	2.79×10^{-39}
Magnesium hydroxide	$Mg(OH)_2$	5.61×10^{-12}
Nickel hydroxide	$Ni(OH)_2$	5.48×10^{-16}
Zinc hydroxide	$Zn(OH)_2$	3×10^{-17}

Further, we can make an estimate of the pH at which precipitation each of these hydroxide species will occur when the OH^- ions are added.[22] For example, if we start with a 1M solution of aluminium nitrate, $Al(NO_3)_3$, the Al^{3+} starting concentration is 1.0 and so precipitation will start when the OH^- concentration is such that $K_s = 3 \times 10^{-34} = a_{Al^{3+}} \cdot (a_{OH^-})^3$. In other words, $a_{OH^-} = 6.7 \times 10^{-12}$. In water, $K_w = a_{H^+} \cdot a_{OH^-} = 10^{-14}$ (i.e. pH + pOH = 14) and so this OH^- ion concentration corresponds to a pH of about 2.83.

If we now consider the case for a 1M solution of $Ni(OH)_2$, we find that the pH at which precipitation will occur is about 6.37. In other words, if hydroxide ions are added to a mixture of Ni and Al ions, the aluminium species should precipitate at a much lower pH than should the nickel species. As we shall see in the following section, this is not the case.

4.4.2.2 Precipitation of several species together: coprecipitation

When two cations are present together in the precipitation solution and the pH of the solution is gradually increased by the addition of an alkali,[23] one would expect the two ions to precipitate separately at different values of the pH corresponding to the individual solubility products; for example, with a solution containing nickel and aluminium ions, one would expect, as discussed above, that the aluminium species precipitate at a pH of ≤ 3 and that the nickel then precipitates at a pH of above 6. What appears to happen in practice is that the aluminium begins to precipitate at low pH but it then re-dissolves and coprecipitates together with the nickel at pH values well below 6. What then is formed is a mixed layered hydroxide compound (with a "brucite" structure) having a structure (see below) corresponding to that of minerals that are members of the so-called hydrotalcite group; in the case of the Ni–Al system, the naturally occurring mineral is called "takovite". Coprecipitated Ni–Al materials were first introduced by British Gas as catalysts for the steam reforming of naphthas to

[22] In these calculations, we assume that the concentrations and activities are approximately the same. For more precise calculations, we would need to know the activity coefficients of the different species.

[23] This has been termed "the rising pH method" of coprecipitation.

produce a methane-rich gas for heating purposes. In the early work on these Ni materials, the formation of the layered structures was not recognized and it was only in a series of patents from BASF in 1970 that their importance was reported.[24] In work that we published in 1975, without knowledge of these patents, we showed that there was some form of interaction between the Ni and Al ions in the precipitates.[25] In a later detailed study of the coprecipitation of Ni/Al catalyst precursors and of the properties of the resultant materials, we showed[26] that the composition of the layered structure depends very much on the way in which the material is prepared. In particular, a much more clearly defined material could be obtained by carrying out the precipitation at constant pH: by adding a solution of the nitrates to water at the same time as adding alkali, maintaining the flows so that the pH remains constant. If the pH of precipitation is 5.0, the resultant precipitate contains nitrate ions whereas, if the pH is above 6.0, carbonate species become predominant. A relatively wide range of composition is possible but much of the work reported in the literature deals with a Ni/Al ratio of 3.0. The layered compound containing Ni and Al species in this ratio and prepared at a pH of 10.0 has the composition:

$$Ni_6Al_2(OH)_{12}CO_3 \cdot 5H_2O$$

If the pH is lower, then the CO_3^{2-} ions are replaced by two NO_3^- ions and the following composition is obtained:

$$Ni_6Al_2(OH)_{12}(NO_3)_2 \cdot 3H_2O$$

The nitrate or carbonate ions balance the excess charge introduced into the structure by the Al^{3+} ions. The structure of takovite is shown schematically in Figure 4.9 in which the preparation and subsequent calcination and reduction of the resultant materials are also depicted. The Ni and Al ions are randomly distributed in an octahedral environment surrounded by OH^- ions in a brucite-like layered structure (brucite is $Mg(OH)_2$) and the brucite layers are stacked one above another with interlayers that are occupied by the charge balancing carbonate or nitrate ions plus the water of crystallization. When two nitrate ions are present in place of one carbonate ion, there is less space over for the water; the interlayer spacing is also dependent on the charge balancing anion, it being greater with the nitrate ions (0.85 nm) than with the

[24] F.J. Brocker, L. Kainer, German Patent 2,024,282 (1970), to BASF AG, and UK Patent No. 1,342,020 (1971), to BASF AG.

[25] T. Beecroft, A.W. Miller, J.R.H. Ross, The use of differential scanning calorimetry in catalyst studies. The methanation of carbon monoxide over nickel/alumina catalysts, J. Catal., 40 (1975) 281.

[26] Some of the work carried out in a collaborative effort between the University of Bradford (UK) and the Technical University of Delft is summarized in the following publications and they can be consulted for further details: G. van Veen, E.C. Kruissink, E.B.M. Doesburg, J.R.H. Ross, L.L. van Reijen, Rn. Kinet. Catal. Lett. 9 (1978) 143; E.C. Kruissink, L.E. Alzamora, S. Orr, E.B.M. Doesburg, L.L. van Reijen, J.R.H. Ross, G. van Veen, in: B. Delmon, P. Grange, P. Jacobs, G. Poncelet (Eds.), Preparation of Catalysts, II (1979) 143; E.C. Kruissink, L.L. van Reijen, J.R.H. Ross, J. Chem. Soc., Faraday Trans. I 77 (1981) 649; L.E. Alzamora, J.R.H. Ross, E.C. Kruissink, L.L. van Reijen, J. Chem. Soc., Faraday Trans. I 77(1981) 665.

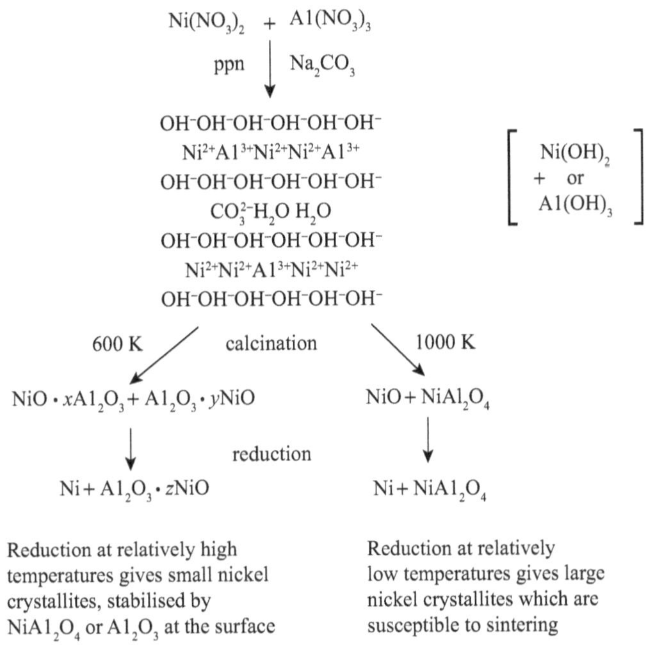

Figure 4.9:
Steps in the preparation of coprecipitated nickel-alumina catalysts. *Source: J. Chem. Soc., Faraday Trans. I, 77 (1981) 665–681. Reproduced with kind permission of the Royal Society of Chemistry.*

carbonate ion (0.77 nm). It is worth noting at this stage that a wide variety of materials with hydrotalcite structures can be prepared (with cations that do not differ greatly in ionic radii from that of Mg^{2+}) and that many of these can be used as catalyst precursors, examples being the Cu-Zn-Al system (see Table 4.3). Further, a variety of different charge-balancing anions can be placed in the interlayer spacing.[27]

One of the greatest advantages offered by these coprecipitated materials containing the hydrotalcite structure is the ease and reproducibility of preparation: as long as the material is made using the same set of conditions and it is washed thoroughly before further treatment to remove traces of undesirable alkali metal species introduced during preparation, the properties are identical from batch to batch. It should be noted that even when carbonate ions are carefully excluded during the preparation, the precipitate often contains at least some level of carbonate species as CO_2 is very easily picked up from the atmosphere. Removal of alkali ions is greatly facilitated if the precipitate is dried at ca. 100 °C to break down its gel structure prior to washing.

[27] By precipitating a solution of Co and Al nitrates using ammonium molybdate, it was possible to make a layered structure in which the molybdate anion was located in the interlayer and the Co and Al ions were in the brucite structure. On calcination and subsequent sulfidation, this material had equivalent behaviour for hydrodesulfurization as did a commercial $CoMoAl_2O_3$ catalyst.

Table 4.3: Some examples of hydrotalcite catalyst precursors prepared by coprecipitation.

M^{2+}	M^{3+}	Balancing anion
Mg^{2+}	Al^{3+}	CO_3^{2-}, OH^-, NO_3^-, etc.
Ni^{2+}	Al^{3+}	CO_3^{2-}, SO_4^{2-}, $V_{10}O_{28}^{6-}$, etc.
Mg^{2+}	Al^{3+}	$V_{10}O_{28}^{6-}$, $Mo_7O_{24}^{6-}$
Fe^{2+}	Al^{3+}	CO_3^{2-}
Co^{2+}	A^{3+}	CO_3^{2-}
Cu^{2+}	Al^{3+}	CO_3^{2-}
Mg^{2+}	Fe^{3+}	CO_3^{2-}
Cu^{2+}, Zn^{2+}	Al^{3+}	CO_3^{2-}
Cu^{2+}, Co^{2+}	Al^{3+}	CO_3^{2-}
Cu^{2+}, Zn^{2+}	Cr^{3+}	CO_3^{2-}
Cu^{2+}, Co^{2+}	Cr^{3+}	CO_3^{2-}

Source: F. Cavani, F. Trifirò, A.Vaccari, Hydrotalcite-type anionic clays: preparation, properties and applications, Catal. Today 11 (1991) 173.

If the coprecipitated catalyst is to be used in a commercial reactor, it must be formed into pellets in the same way that the supports used in supported catalysts must be formed. For the Ni-Al materials discussed above, this is usually done by adding a lubricant of some sort (e.g. graphite) to the precipitate prior to calcination and pelleting the resultant mixture in a commercial tabletting machine. It is then calcined to remove the hydroxyl ions and carbonate species from the structure and, before use, it is reduced in a flow containing hydrogen in the same way as for a supported catalyst (see Figure 4.9). If the calcination is carried out at a relatively low temperature of 350–500 °C), the resultant material contains small particles of NiO stabilized by the presence of some Al^{3+} ions together with an alumina-rich phase containing trace quantities of Ni^{2+} ions; if the calcination is at higher temperature, the calcined material contains phases of NiO and $NiAl_2O_4$. Reduction of the material calcined at low temperature gives small Ni particles and these are stabilized by the alumina on the surface (see Figure 4.10); the material calcined at higher temperatures contains large Ni particles, which are not so well stabilized.

TASK 4.8 Catalysts Derived from Hydrotalcites

Using the paper by Cavani et al. (see Table 4.3), examine the literature in relation to one or more of the uses of hydrotalcite precursors. (It should be noted that this paper was published before Scopus was created and so the linkages are not as good as for recent papers.) Various approaches to the vast literature on hydrotalcites could be used. For example, you can move forward in time (e.g. by following a highly cited author who has cited the Cavani review) to examine more recent catalytic work using these materials. Alternatively, you could concentrate on finding whether there have been any advances in structural knowledge or in the range of compounds that have been examined.

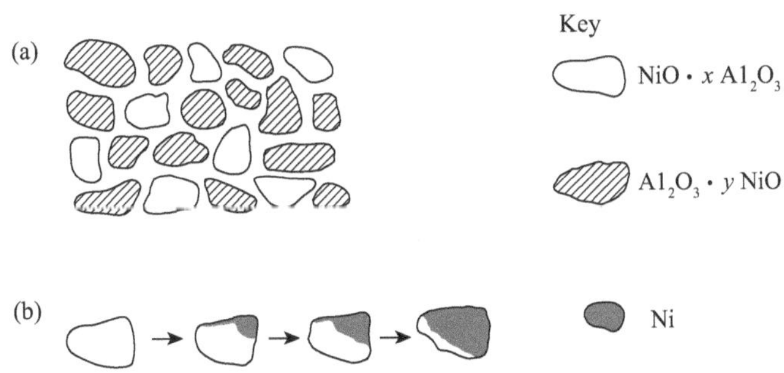

Figure 4.10:
Model for the calcined Ni-Al catalyst prepared by coprecipitation and its reduction. (a) Calcined material; (b) progression of reduction. The NiO particle size depends on the calcination temperature. The reduction rate is limited by the growth of a Ni nucleus in a NiO particle of constant size; sintering may occur after completion of the reduction by agglomeration of the Ni crystallites but is hindered by the presence of alumina on the surface of these particles. *Source: L.E. Alzamora, J.R.H. Ross, E.C. Kruissink and L.L. van Reijen, J. Chem. Soc., Faraday Trans. I, 77 (1981) 665–681. Reproduced with kind permission of the Royal Society of Chemistry.*

4.4.2.3 Precipitation and coprecipitation using urea decomposition

Urea decomposes in aqueous solution at temperatures of about 90 °C to give NH_4^+ and OH^- ions. The method of preparation termed "deposition precipitation" was first developed by Geus et al.[28] and involves adding an aqueous solution of a salt of the metal to be deposited (typically $Ni(NO_3)_2$) and urea to a support such as silica. The suspension is then heated and the urea hydrolyses, giving OH^- ions evenly throughout the solution. The nickel species then precipitate homogeneously on the support surface. This is followed by calcination and reduction as required.[29]

Urea decomposition has also been used successfully[30] to prepare coprecipitated hydrotalcite structures within the pores of a preformed support such as α-Al_2O_3. In this case, the urea solution contained not only $Ni(NO_3)_2$ but also $Al(NO_3)_3$ and $La(NO_3)_3$ and it was vacuum impregnated into the macro-pores of the α-Al_2O_3; the excess solution was then removed prior to heating the support. Precipitation of the layered material (in which the La was also incorporated in the brucite structure) was thus achieved; if necessary, the procedure could be repeated several times in order to increase the total Ni loading by partially decomposing the

---

[28] J.A. van Dillen, J.W. Geus, L.A.M. Hermans, J. van der Meijden, in: G.C. Bond, P.B. Wells and F.C. Tompkins (Eds.), Proceedings of the 6th International Congress on Catalysis, Chem. Soc. 2 (1977) 677.

[29] Such precipitation onto a support has also been achieved simply by adding an alkali drop-wise to a suspension of the support in the salt solution with very vigorous stirring; however, this process is much more difficult to control and generally results in uneven distribution of the active components, there being a tendency to accumulate more on the external surfaces of the support rather than within the pores. Hence, deposition precipitation is more reliable.

[30] K.B. Mok, J.R.H. Ross, R.M. Sambrook, Preparation of Catalysts III, in: G. Poncelet, P. Grange, P.A. Jacobs (Eds.), Stud. Surf. Sci. Catal., 16 (1983) 291.

precipitate prior to each impregnation/deposition step. Ni contents of up to 8 wt% Ni were achieved. The addition of La was found to increase the Ni dispersion in the resultant catalysts and to stabilize the structure of the calcined and reduced catalysts as well as to improve the stability of the catalysts against carbon deposition during use in steam reforming.

TASK 4.9 Deposition Precipitation

Use Scopus or Web of Science to search the literature for papers on "deposition precipitation". Among the papers you locate, you will find a paper by P. Burattin, M. Che, C. Louis (J. Phys. Chem. B, 102 (1998) 2722) on the deposition of Ni on silica. Use this paper as a source article to search for other papers on deposition precipitation. For example, you will find that K.P. de Jong has cited this paper in his article on the synthesis of supported catalysts in Curr. Opin. Solid State Mater. Sci., 4 (1999) 55. These two articles will provide you with sources (including the paper by Van Dillen et al. referred to above) that consider the use of deposition precipitation for the preparation of various different sorts of catalyst. Possible topics for detailed study might be supported nickel or gold catalysts; however, many other topics might also be selected.

TASK 4.10 Copper-Containing Catalysts for Methanol Reforming

Breen and Ross* have examined the behaviour of a series of Cu—Zn—Al—Zr catalysts prepared by a sequential precipitation technique and have studied the effect of method of preparation and pretreatment on the total and Cu metal areas as well as on the activity for the steam reforming of methanol. Study this paper to learn more about the preparation and characterization methods used, especially the N_2O decomposition technique for measuring Cu surface area. You will also be able to see the effect of promoters on the activities and stabilities of the Cu catalysts. Where necessary, study some of the references given as well as carrying out a Scopus or Web of Science search on the paper to identify important recent papers published on the same subject.

* J.P. Breen, J.R.H. Ross, Methanol reforming for fuel-cell applications: development of zirconia-containing Cu-Zn-Al catalysts, Catal. Today 51 (1999) 521—533.

4.4.2.4 Decomposition of amino complexes

Schaper et al. have reported an alternative to homogeneous precipitation for the preparation of the hydrotalcite structures discussed above.[31] They prepared the amino complex of Ni by adding concentrated ammonia to a solution of $Ni(NO_3)_2$ containing concentrated nitric acid and then adding sodium hydroxide solution until the pH was 10.5. The appropriate alumina support, in the form of extrudates, was then added to the solution and heated to 90 °C while

[31] H. Schaper, E.B.M. Doesburg, J.M.C. Quartel and L.L. van Reijen, Synthesis of methanation catalysts by deposition-precipitation, in: G. Poncelet, P. Grange, P.A. Jacobs (Eds.), Preparation of Catalysts III, Stud. Surf. Sci. Catal. 16 (1983) 301.

bubbling CO_2 through the solution. The pH changed rapidly to 7.5 and then dropped further by a slower process. It was found that some of the alumina dissolved and then re-precipitated, together with the Ni, as the hydrotalcite compound. After calcination and reduction, the resultant materials had similar properties to those of those made by the more conventional coprecipitation route discussed above.

4.5 Catalyst Characterization

Having prepared a catalyst, it is essential to be able to characterize it. The main purpose of catalyst characterization is to be able to define those characteristics of the catalyst that has been prepared that are responsible for, or are related to, its catalytic behaviour. A secondary purpose is to enable reproduction of any preparation carried out, enabling one to show that each sample prepared under the same conditions and with the same chemical composition has the same physical and chemical (including catalytic) properties. The techniques used for characterization range from the straightforward determination of the all-over chemical composition and phase structure to details of the active species present on the catalyst surface. The following paragraphs are focused predominantly on the more routine methods used in the majority of laboratories examining catalytic properties. However, some attention is also given to some of the more advanced specialist methods available to the catalytic chemist.

4.5.1 Chemical Composition

The chemical composition of a catalyst is a critical parameter and should always be given in any description of the properties of a particular sample. While the composition is often exactly the same as that which was aimed for during the preparation of the sample and can be calculated from the concentrations of the reagents used in the preparation, this is not always the case. For example, samples prepared using the impregnation method may have metal ion loadings below those calculated by assuming that all the added species were adsorbed during preparation, this being a result of incomplete adsorption. Similarly, not all the ionic species present in solution during a coprecipitation procedure may be precipitated or some may be re-dissolved during a washing step. Hence, the chemical composition should be obtained to make sure that the sample composition is as expected.

The chemical composition of a catalyst can be obtained by conventional analytical methods: dissolving the sample in a suitable solvent (often an acid) and then carrying out standard wet analytical procedures (either volumetric or gravimetric) to obtain appropriate figures. However, such methods have a variety of problems; for example, the various components of a multicomponent catalyst can cause interference with one another during analysis or one or other component may not be completely soluble under the dissolution conditions applied. Spectroscopic methods are also frequently used to determine the composition of the solution resulting from the dissolution process, the most commonly used being atomic emission spectroscopy (AES). Many other spectroscopic methods, for example UV-visible spectroscopy, have also been used for analysis purposes for particular catalyst compositions, each having particular advantages for a particular species, and it is advisable to study the relevant literature for any type of catalyst to find which method is most commonly used for that composition.

Another common method applied to obtain the elemental composition of a catalyst is X-ray fluorescence (XRF). A useful tutorial on the technique is given at: http://www.horiba.com/scientific/products/x-ray-fluorescence-analysis/tutorial/. A more general discussion of the method is available at: http://en.wikipedia.org/wiki/X-ray_fluorescence/. In order to carry out an analysis, a sample of the catalyst is powdered and formed into a glass with, for example, sodium borate, in order to give a disc with a flat surface that can be inserted in a standard XRF instrument. The resultant peaks are compared with those of standard mixtures of the various components prepared in the same way to enable the composition to be determined. The AES technique cannot be applied to hydrogen, helium and lithium ($Z \leq 3$) and is best if applied for elements heavier than sodium. Also, because the escape depth of secondary X-rays form a sample containing the lighter elements is relatively short, the sample must be quite uniform and the disc must be flat.

Other physical techniques that can be used to give the chemical composition of a catalyst include X-ray photoemission spectroscopy (XPS), auger spectroscopy, secondary ion mass spectrometry, Rutherford backscattering spectrometry, low energy ion scattering (LEIS), electron microscopy (measuring backscattered X-rays in a technique known as energy dispersive analysis of X-rays, EDAX) and extended X-ray absorption fine structure (EXAFS).[32] As many of these techniques are surface sensitive (in other words, the signals measured emanate from the surface or from near the surface), the results obtained are related more to surface than to bulk composition. Of these, LEIS is probably the most surface sensitive.

TASK 4.11 LEIS Applied to Supported Catalysts

R.H.H. Smits and his colleagues (R.H.H. Smits, K. Seshan, J.R.H. Ross, L.C.A. van den Ootelaar, J.H.J.M. Helwegen, M.R. Anantharaman, H.H. Brongersma, A low energy ion scattering (LEIS) study of the influence of the vanadium concentration on the activity of vanadium-niobium oxide catalysts for the oxidative dehydrogenation of propane,J. Catal., 157 (1995) 584–591) have compared LEIS and XPS results for coprecipitated vanadia-niobia catalysts and have shown that LEIS is the more surface-sensitive technique. Use this reference as a starting point to examine the current status of the LEIS method, especially in relation to studying supported catalysts.

[32] These and other techniques are admirably described in the book by J.W. Niemanstverdriet entitled *Spectroscopy in Catalysis. An Introduction*, Wiley VCH, Weinheim, Germany, First Edition 1993, updated regularly since.

4.5.2 Phase Structure

Having determined the chemical composition of a catalyst, it is also often necessary to learn more about the phase structure. This is generally achieved using X-ray spectroscopy (http://en.wikipedia.org/wiki/X-ray crystallography), a standard method which can be used to give the structures of the phases present in the catalyst.

TASK 4.12 X-ray Diffraction to Study Catalyst Structures

An excellent example of the application of XRD to the study of catalyst structures is a review article by D.C. Puxley, I.J. Kitchener, C. Komodromos and N.D. Parkyns (The effect of preparation method upon the structure, stability and metal/support interactions in nickel/alumina catalysts, in: G. Poncelet, P. Grange, P.A. Jacobs (Eds.), Preparation of Catalysts III, Stud. Surf. Sci. Catal. 16 (1983) 237–271). Identify other papers which concentrate on the use of XRD, in particular, those using in-situ methods.

4.5.3 Physical Texture

The physical texture of a catalyst or its precursor(s) is often examined using scanning electron microscopy (SEM). (The associated EDAX measurements, see above, can be used to analyse the composition of selected areas of the image). Because the resolution of SEM is about 5 nm, the method does not give information on an atomic scale but can give useful data on the form of micropores and, for example, changes occurring during pelletting of a catalysts sample. The much higher resolution obtainable using transmission electron microscopy (TEM) allows one to observe even individual atoms of a surface, this being particularly useful when studying, for example, small metal crystallites on a particular support. Careful and systematic measurements using TEM also allow estimates to be made of the particle size distributions occurring in a particular sample. Detailed information on such crystallites can also be obtained using scanning probe microscopic methods such as atomic force microscopy (AFM) and scanning tunnelling microscopy (STM).[33]

The technique of X-ray line broadening can be used to give information on particle sizes present in the catalyst sample. This method uses the Scherrer equation:

$$\mathrm{B}(2\theta) = \frac{K\lambda}{L\ \cos\theta}$$

where the peak width, $B(2\theta)$, at a particular value of 2θ (θ being the diffraction angle, λ the X-ray wavelength) is inversely proportional to the crystallite size L; the constant K is

[33] It should be recognized that TEM, AFM and STM sample only relatively small areas of a catalyst and so many parts of a catalyst surface must be examined to obtain fully representative results.

a function of the crystallite shape but it is generally taken as being about 1.0 for spherical particles.[34] X-ray line broadening gives average values of crystallite size which in most cases agree quite closely with those obtained by TEM; however, the problems of obtaining representative results are less exacting.

4.5.4 Catalytic Activity, Selectivity and Stability Under Operating Conditions

The most important properties of a catalyst are its activity and selectivity under operating conditions. Hence, an essential feature of characterizing a newly synthesized catalyst is the determination of the catalytic properties of the new material. It is hardly necessary to say that there is no sense in making a new catalyst if it does not exhibit good catalytic properties. Nevertheless, it is amazing to find the number of publications that do not give adequate data on the properties of the samples under investigation, particularly on the relation of the behaviour of the new material for the reaction of interest relative to that of known catalysts. Further, in making comparisons, it is essential to have a full understanding of the effects of differing testing conditions between one laboratory and another on the results obtained. Laboratory methods of testing catalysts from the point of view of characterization and comparison will therefore be the topic of Chapter 5. In order to make meaningful comparisons, it is often essential to know the kinetic expression describing the reaction in question over the new catalyst. Chapter 6, therefore, considers the kinetics of catalytic reactions and discusses some of the more commonly encountered mechanistic approaches.

[34] The diffraction of electrons and neutrons can also be used for the same purposes but these methods are much less frequently encountered, probably because X-ray diffraction is so generally available.

CHAPTER 5

Catalytic Reactors and the Measurement of Catalytic Kinetics

Chapter Outline

5.1 Introduction

It was shown in Chapter 3 that a catalyst speeds up a reaction by offering an alternative, more energetically favourable route for the reaction. We have also seen in Chapter 4 how a catalyst can be made and characterized. However, an important aspect of catalyst characterization was omitted from Chapter 4: how to examine the behaviour of a catalyst for the reaction of choice. This chapter, therefore, answers the question: how does one measure the rate of a catalytic reaction in a laboratory and compare one catalyst with another? The problems of scale-up to a full-scale reactor will be discussed in the next chapter.

It is assumed that the reader is conversant with the methods of studying the kinetics of homogeneous reactions and also with the concepts of order of reaction and the effect of temperature on reaction rate (the Arrhenius equation). These are well covered in many texts on Physical Chemistry and one of these should if necessary be perused prior to studying this

Heterogeneous Catalysis. DOI: 10.1016/B978-0-444-53363-0.10005-2

chapter. In studying the kinetics of a homogeneous liquid phase reaction such as the hydrolysis of an ester by an acid, one needs to be able to measure the concentration of at least one of the reactants or products as a function of time at a specific temperature. This is most commonly done in a vessel such as a stirred beaker immersed in a suitable thermostat, a typical constant volume reactor, from which small aliquots of the reaction mixture can be withdrawn for analysis (e.g. by titration) at regular intervals. The data are then plotted as a graph of concentration of reactant or product as a function of reaction time. In order to determine the kinetic expression governing the reaction, the data are fitted to various integrated rate expressions and rate constants are determined; the activation energy for the reaction is then calculated (see Box 5.1).

A similar approach would be adopted if one was to follow the kinetics of a gas-phase reaction such as the oxidation of NO to give NO_2:

$$NO\,(g) + 0.5\,O_2 \rightarrow NO_2(g) \tag{1}$$

in which (g) indicates that the reactants and product are in the gas phase. In this case, an analytical method has to be used which would allow the partial pressures of the reactants and products to be measured; if the stoichiometry of the reaction is followed exactly (e.g. neither N_2O nor N_2O_5 is formed), a simple pressure change measurement using a constant volume reaction system suffices, for example, a spherical pyrex vessel carefully thermostated in

BOX 5.1 Kinetics of a Homogeneous Reaction

Consider the reaction in which the starting concentrations of A and B are the same:

$$A + B \rightarrow C + D$$

It is assumed that the stoichiometry of the reaction is given by the chemical equation and so the rate of reaction can be expressed by:

$$\text{Rate} = -k_T \cdot [A]^n (= -k_T \cdot [B]^n)$$

where k_T is the rate constant at temperature T (in K) and n is the order of the reaction. If the reaction is first order, $n = 1$ and the equation may be integrated to give:

$$\ln([A]/[Ao]) = k_T \cdot t$$

where [Ao] is the initial concentration of A and t is the time of reaction. Hence, a plot of ln [A] vs. t gives a straight line of gradient $-k_T$. The experiment can be repeated at various temperatures, T, and the values of k_T can be determined for each temperature.

As the rate constant is related to the activation energy, E_r, by the Arrhenius equation:

$$k_T = A\exp(-E_r/RT)$$

a plot of ln k_T vs. $1/T$ yields a straight line of gradient $-E_r/R$, where R is the gas constant.

a suitable container.[1] However, many other analytical methods, for example, gas chromatography or mass spectrometry, can be used to monitor the partial pressures of the individual reactants and products.

If a fast reaction is being studied, then it is not practicable to use a constant volume system of the type described above since insufficient analyses can be carried out through the duration of an experiment. A wide range of techniques have therefore been developed for studying the kinetics of fast reactions, these including flow reactors using spectrophotometric measurement of the concentrations of reactant and product. In such reactors, reactants are brought together rapidly and thoroughly mixed at the entrance to a flow tube. The concentrations of reactants and products are then measured at various distances from the point of mixing; these corresponding to the time of reaction. Alternatively, in the so-called stop-flow reactor, the flow can be stopped at a given moment and the concentrations measured as a function of time at a particular position downstream of the point of mixing.

5.2 Static Reactors

By analogy with the reactors used for homogeneous systems discussed above, it is possible when investigating a catalytic reaction occurring at the gas-solid interface to inject the reaction mixture into a constant reaction volume containing a sample of the catalyst (Figure 5.1) and to measure the concentrations of the reactants and products as a function of time of reaction. If such an approach is adopted, it is important to recognize an important factor in studying the rates of such reactions: the rates at which gas-phase mixing of reactants and products can take

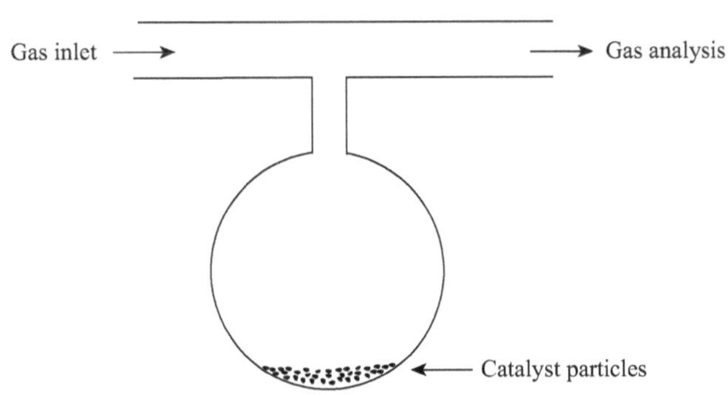

Figure 5.1:
Low pressure constant volume reactor used for the study of a catalytic reaction on a catalyst powder or granules.

[1] For such measurements at higher temperatures, the reactor may be immersed in a heated liquid or in a fluidized sand bath or furnace.

place, this in turn depending on the *mean free paths* of the constituent molecules. The mean free path, λ, of a molecule is given by:

$$\lambda = RT/\left\{\sqrt{2}\cdot N_A\cdot\sigma\cdot p\right\} \quad (2)$$

where R is the gas constant, T is the temperature (K), N_A is Avogadro's number, σ is the collision cross-section of the molecule[2] and p is the pressure. For oxygen at room temperature, the mean free path at a pressure of one atmosphere can be calculated to be 7.3×10^{-8} m = 73 nm. Thus, it is clear that there will be many more collisions between the constituent molecules of pure gas or of a gas mixture than with the walls of a reactor that has a diameter of several millimetre or greater. This has several significant consequences. Most important for our present purposes is that because the effective mixing of two or more gases depends on the movement of the molecules past one another and that this movement will be hindered by collisions between the molecules, it will be difficult to maintain perfect mixing in a constant volume reactor; this is because the reaction takes place only in one part of the reactor – at the surface of a catalyst bed located at the base of the reactor – rather than throughout the reactor as would be the case in a homogeneous reaction. This means that if a catalytic reaction is carried out on a small sample of catalyst in such a reactor at a pressure of 1 bar and analysis is carried out using, for example, mass spectrometry, the spectrometer entrance being located some distance from the catalyst sample, the lack of mixing can lead to spurious results.[3]

A solution to the problem of mixing is to carry out the reaction at lower pressures, for example, in a pre-evacuated reaction volume, introducing the reactant gases via suitable valves. As a rough rule of thumb, if the mean free path approximates to the dimensions of the reactor and the numbers of collisions with the walls becomes significant relative to the number of gas-phase collisions, perfect or near-perfect mixing can be expected. In practice, this means that the upper pressure of the reactor should not exceed about 10^{-3} bar (mean free path approximately 10^{-4} m, 0.1 mm) if mixing rates are to be sufficiently high to avoid analytical problems. Box 5.2 gives an example of an investigation carried out using such a reactor system: the steam reforming of methane over a pre-reduced Ni/alumina catalyst. The use of a low-pressure system has two advantages. Firstly, the catalyst can be pre-reduced in situ and the reduction process can be monitored easily; and, since the steam reforming of methane is an endothermic equilibrium reaction, the conversion approaches 100% at lower temperatures (ca. 600 °C) than if the reaction was to be studied at higher pressure; at higher pressures, temperatures above 800 °C are required to give complete conversion.

2 The value of σ depends on the molecular weight of the molecule in question; for hydrogen, the value is 0.27 nm^2, while for oxygen it is 0.40 nm^2 and for benzene it is 0.88 nm^2.

3 Similar arguments regarding the importance of the mean free path in relation to mixing of gases apply also to the situations encountered with rate determinations with catalyst pellets: in this case, limitations may be due to internal and external diffusion. We have already referred to these in relation to Figure 4.1 and will return to the subject again later. Another way of avoiding diffusion limitations is to use micro-channel reactors as will be discussed in Section 5.10.

BOX 5.2

As an example of the use of a reactor operated at low pressure, we will consider work carried out using a low-pressure system with mass-spectrometric analysis in which the steam reforming of methane over Ni/Al_2O_3 catalysts was examined[a]:

$$CH_4 + H_2O \rightarrow CO + 3H_2 \quad (i)$$

Analysis of the reaction mixture was carried out using a small continuously evacuated mass-spectrometric "residual gas analyser" attached to the constant volume reaction system by a capillary leak. This leak was situated some centimetres from the reactor. To improve gas circulation and mixing, a small centrifugal glass pump was included in the system and this allowed the system to be used as small gas-recirculation reactor (see Section 5.3). It was found that when a dose of one gas was added to a gas already in the reactor, effectively instantaneous mixing of the two gases could be attained. As a consequence, the reactor could be operated using a shallow bed of powdered catalyst and kinetic data could be obtained which were effectively free of problems of internal or external diffusion limitations.

It was found that the rate of reaction of methane with a 75 wt% Ni/Al_2O_3 catalyst was given by the following equation:

$$-dn/dt = 1.78 \times 10^{18} P_{CH_4}^{1.0} P_{H_2O}^{-0.5} \exp\left[\frac{-29000}{RT}\right] \quad (ii)$$

Assuming that the reaction adheres to Eq. (i) under the conditions used, for equimolar initial concentration of methane and water the rate of reaction can be re-written as:

$$dP_{CH_4}/dt = kP_{CH_4}^{0.5} \quad (iii)$$

Integration gives:

$$P_{CH_4}^{0.5} = \text{const.} - kt/2 \quad (iv)$$

Hence, plots of the square root of the methane pressure vs. time should give straight lines. Figures (a)–(d) show such plots calculated from data obtained at 873 K. Figure (a) shows the results of a number of different experiments and demonstrates the reproducibility of the results. It also shows that straight-line behaviour is obtained for times of reaction greater than about 50 s. It would appear that the initial deviation from straight-line behaviour is due to the fact that the initial stages of reaction occur with the catalyst under oxidizing conditions but that once reducing conditions are obtained, the expected behaviour results.[b] Figure (b) shows the effect of adding hydrogen to the feed. The major effect seen is that the addition of increasing amounts of hydrogen causes a reduction of the initial deviation from linear behaviour. Finally, Figures (c) and (d) show that both CO and CO_2 have inhibiting effects on the reaction, presumably due to competition by these molecules for the active Ni (or NiO) sites.

(Continued)

BOX 5.2—Cont'd

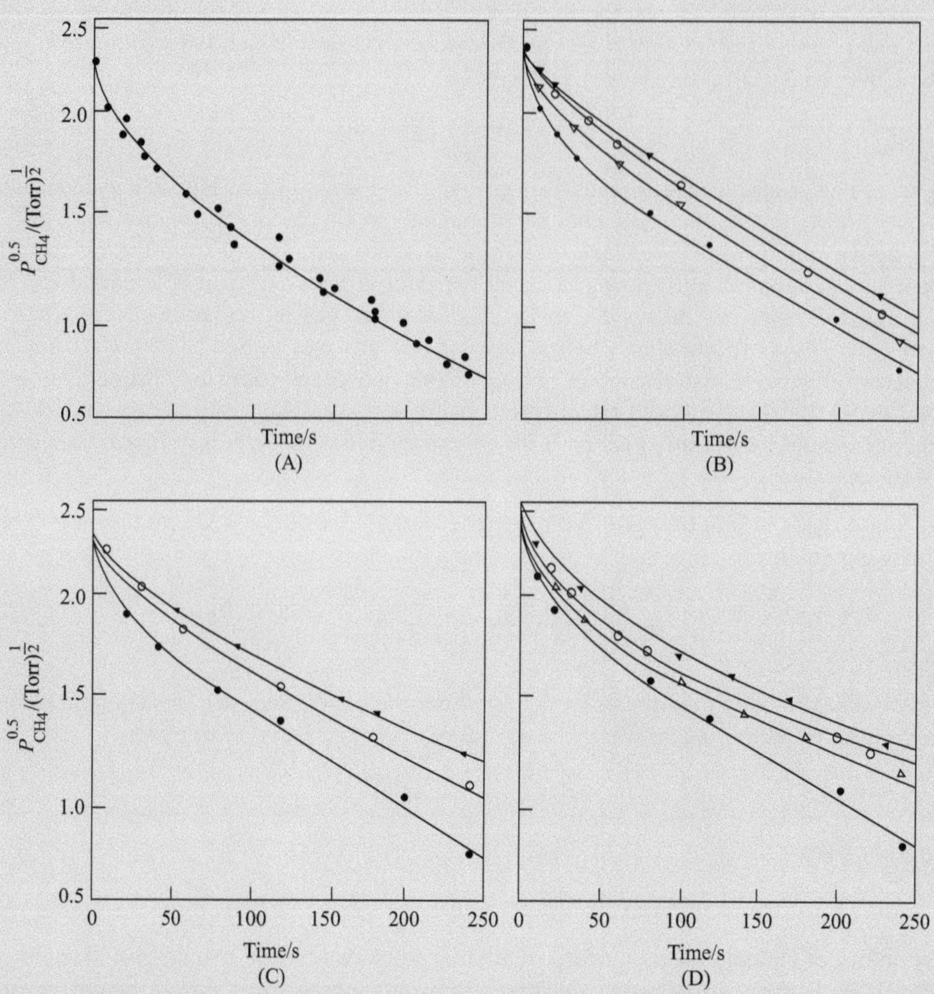

Reproduced with kind permission of the Royal Society of Chemistry.

[a] J.R.H. Ross, M.C.F. Steel, J. Chem. Soc., Faraday Trans. I 69(1973) 10.

[b] The topic of oxidizing/reducing conditions was considered in Chapter 3, Section 3.

5.3 Stirred and Recirculation Reactors

5.3.1 Stirred Reactors

Many of the advantages of the low-pressure constant-volume (static) reactor are also to be found in the use of stirred and recirculation reactors. Stirred reactors are more commonly used for batch reactions without any flow through the reactor and are often used for solid/liquid/gas reactions in which the solid catalyst is suspended in powder form in a liquid which may be an inert or one of the reactants. The concentrations of reactant(s) and product(s) are measured as a function of time and the concentration vs. time of reaction data are handled in the same way as for the constant volume reactor discussed above. In principle, there can also be a flow of reactants into the reactor and of products out of the reactor when the data are handled in the same way as for the stirred recirculation reactor discussed in the next paragraph.

5.3.2 Recirculation Reactors

In recirculation reactors, the reaction mixture is recirculated externally at a rate sufficiently high to ensure that the concentrations of the reactants and products are effectively constant throughout the catalyst bed. In most cases, this is combined with a steady flow of reactant into the reactor and removal of the reaction mixture at another position relative to the bed. The difference between the rates of flow of reactant in and out of the catalyst bed is equivalent to the rate of reaction, as shown schematically in Figure 5.2. As long as the recirculation rate is high enough compared with the rate of reaction, the concentrations of the reactants and products at the exit are the same as those within the bed. A typical recirculation reactor is the Berty Reactor (Figure 5.3), the operation of which is well described in the Autoclave Engineering brochure (see Box 5.3) and another is the Carberry Spinning Catalyst Basket Reactor described in the same brochure. The ideal concentration profile through a recirculation system incorporating a Berty or Carberry Reactor will be the same as shown in Figure 5.2 for a recirculation reactor. On the left-hand (entrance) side of the reactor, the reactant concentration is constant until the point at which the flow enters the reactor. The concentration then drops immediately to the concentration that prevails throughout the reactor and this is the concentration that leaves the reactor at the exit. The product concentration increases just after the entrance to a level that remains the same throughout the reactor and also prevails at the exit. To obtain such a profile, it is necessary that the recirculation rate far exceeds the rate of reaction; in practice, this may not

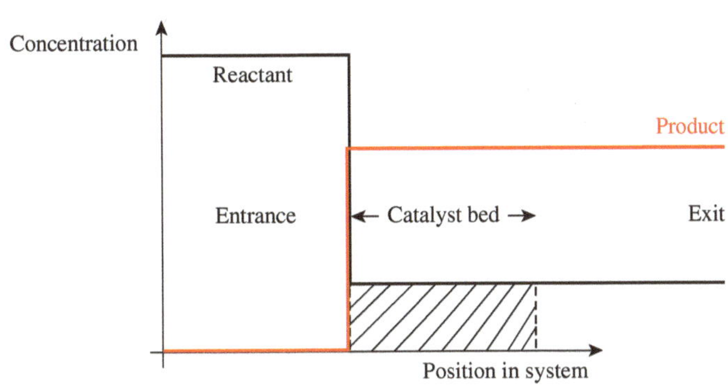

Figure 5.2:
Typical concentration profiles through a recirculation reactor.

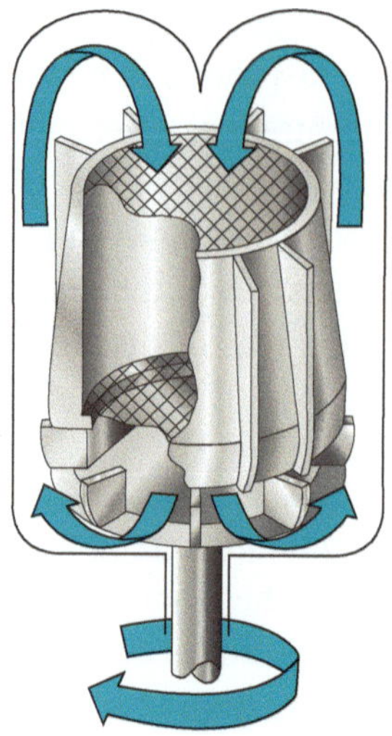

Figure 5.3:
A Berty-type recirculation reactor. (*Courtesy of Autoclave Engineers*).

always be the case and so there may a slight deviation from the ideal profile shown. In practice, it is likely that there will be a second reactant as well as additional products; there will thus be an additional profile for each additional reactant and product but these are omitted for clarity. The recirculation reactor has the advantage over other reactor types that it allows the rate of the reaction to be determined directly from the rate of flow of reactant and the conversion; the conversion is given by the change in concentration divided by the initial concentration.

TASK 5.1

Show how you would carry out an experiment in a Berty reactor to obtain kinetic data such as those represented by Eq. (ii) of Box 5.2.

BOX 5.3

A typical example of a process using a three-phase stirred tank reactor is the hydrogenation of fats and oils using a powdered Ni/Al_2O_3 catalyst. The catalyst is suspended in the liquid to be hydrogenated in a stirred tank and hydrogen is introduced as either a static atmosphere or as a flow. The level of hydrogenation achieved can be determined easily by following the hydrogen consumption. Further information on the hydrogenation process can be obtained by following the Wikipedia link below, while data on the types of reactor used can be found by registering at the Parr website. http://en.wikipedia.org/wiki/Hydrogenation/; http://www.parrinst.com/ Other examples of stirred reactors are found in a document available at the website of Autoclave Engineers (http://www.autoclaveengineers.com/ae_pdfs/CR_Catreact_SelGd.pdf).

5.4 Flow Reactors

By analogy with the flow reactors used to study fast homogeneous reactions (see Section 5.1), most laboratory investigations of catalytic reactions are carried out using flow reactors. In such reactors, the catalyst is placed in a *packed bed* which is normally held between two porous pads or frits. The reactor is usually mounted vertically to make sure that the catalyst bed is packed as uniformly as possible and the gas flow is usually directed from top to bottom to ensure that the packing of the bed is not disturbed by the flow.[4] A typical plug-flow reactor set-up is shown in Figure 5.4. The gas flow entering the reactor is carefully controlled, generally by mass-flow controllers, and the gas exiting the bed is analysed using a suitable system such as a gas chromatograph. The packing of the bed must be such that the gas flows through it in a tortuous fashion (see inset) and this ensures good lateral mixing and that there are no concentration gradients at right angles to the direction of flow. To help ensure good mixing in the inter-particular space, the particles of catalyst must have

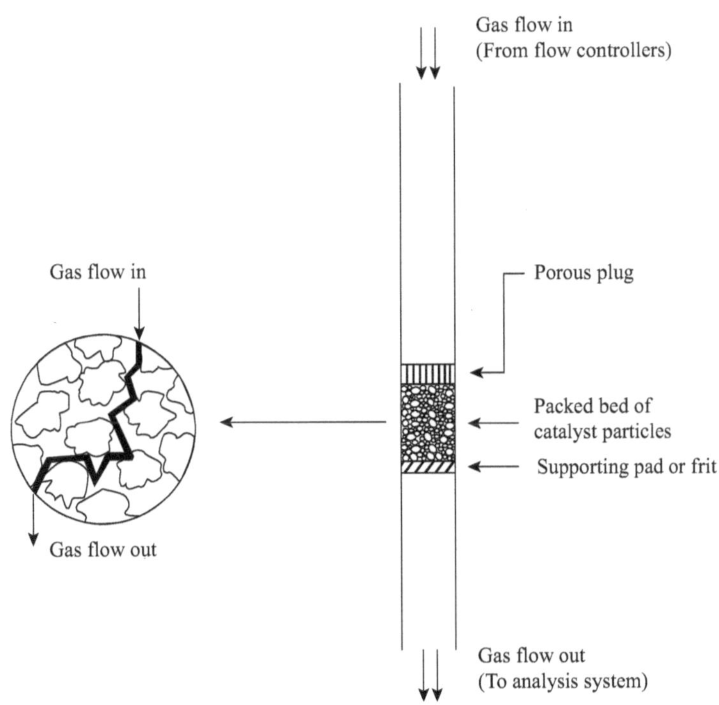

Figure 5.4:
Schematic representation of a catalytic flow reactor. The inset shows the flow through a section of the bed.

---

[4] Horizontal beds have also been used in some circumstances and flow from bottom to top of a vertical bed can also be employed; the latter is always the case if a fluidized bed is employed (see Section 5.5).

BOX 5.4 Catalyst Beds Containing Pellets or Extrudates

Figure 5.4 depicts small irregular particles of catalyst, but it is also possible to construct such a reactor using pellets or extrudates. However, because of the dimensional limitations outlined above, pellets and extrudates generally require larger reactors than those used in most laboratories and the use of such particle dimensions is generally only encountered in pilot-scale or full-scale reactors.

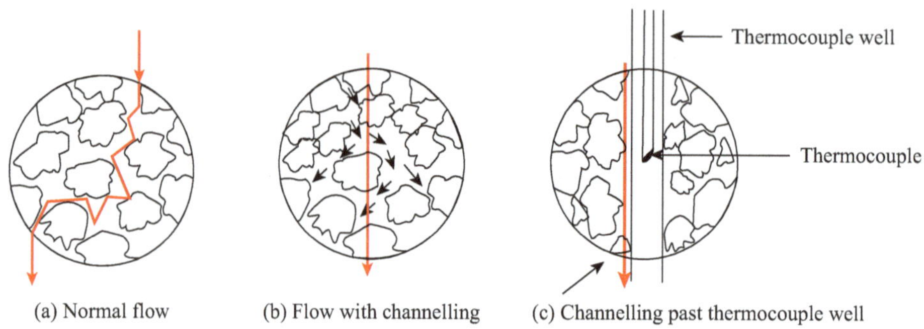

Figure 5.5:
Schematic representation of channelling in a catalyst bed: (a) Normal distribution as in Figure 5.2; (b) channelling due to poor stacking of catalyst; (c) channelling along the side of the thermocouple well.

a diameter, which is significantly smaller than the diameter of the reactor (more than 10 particles across the diameter, preferably more) and the bed depth should be more than at least three times the bed diameter to ensure that there is no back-mixing of the products with the reactants. The reactor is generally placed in a suitable furnace and a thermocouple is inserted either in direct contact with the outside of the reactor next to the bed or in a sleeve inserted in the bed itself. When a thermocouple sleeve is inserted in the bed, it is important to ensure that the flow patterns are not disturbed and that there is no chance of channelling of the reactants past the bed. Figure 5.5 compares schematically (a) the ideal transport through a bed (as in the inset of Figure 5.4), (b) the channelling occurring in a badly packed catalyst bed and (c) and channelling along the wall of a thermocouple well. Another important parameter is that of thermal conductivity of the catalyst bed, something that is most important if we are dealing with a strongly exothermic or a strongly endothermic reaction. It is essential that heat is transferred rapidly between the walls of the reactor and the bed to ensure that there are no lateral temperature gradients. The ideal flow through a packed bed is as shown in Figure 5.6. An even concentration front moves through the bed in such a way that there are no differences in concentration between the centre and the walls, so-called "plug flow".

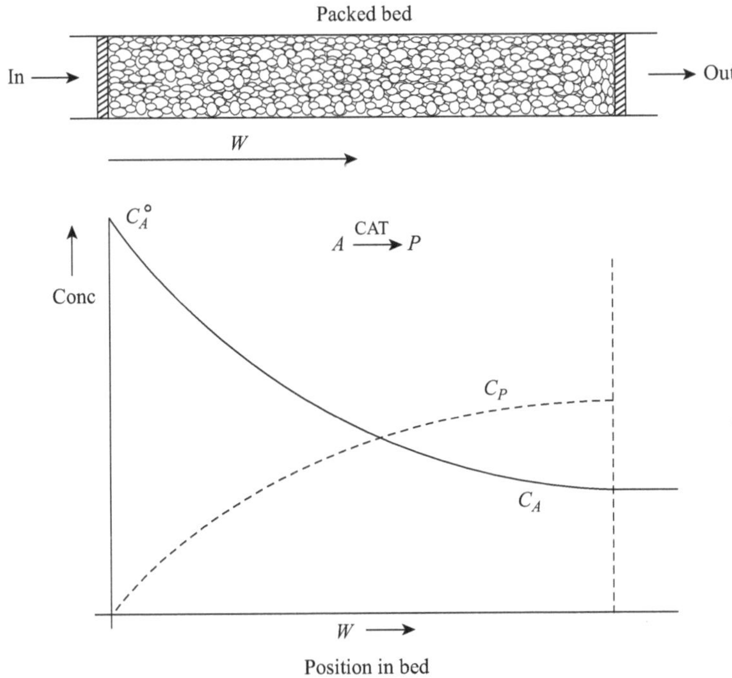

Figure 5.6:
Concentration vs. position in a packed bed for plug flow conditions. The bed is shown as mounted horizontally so that the plot is equivalent to that for concentration vs. time for a homogeneous reaction or a constant volume system (see Box 5.2). Note that the *x* axis is shown as *W*, the weight of catalyst to that point in the bed; if it was expressed as *W*/*F* (*F* = flow rate), this would have the dimensions of time. If the effective density of the bed is 1.0 g cm^3, then *W*/*F* is the time spent in the bed.

If plug flow is achieved, then the concentrations of reactants and products achieved at any position in the bed will be equal across the diameter of the bed and will be a function of the initial concentrations and of the time for which the reactants have been in contact with the bed, as illustrated in Figure 5.6. The contact time, τ, will depend on the rate of flow, the cross-sectional area of the bed and the distance from the entrance. Conventionally, the rate of flow, F, is generally expressed in cm^3(stp)/min (although other units can also be used, for example, molecules/s or mol/h). If the total volume of the bed is V_T, then the time taken for a plug of reacting gas to pass through the bed, τ_T, will be V_T/F, assuming that there is no significant volume change in the flowing gas as a result of reaction.[5] At any position x in the middle of the bed, the time taken for the gas to reach that point, τ, will by analogy be given by V_x/F, where V_x is the volume of catalyst up to that point. The volume occupied by the catalyst will depend on

[5] This is a reasonable assumption if the reactants and products are diluted with a carrier gas. However, if there is no appreciable concentration of diluent, the treatment has to be modified by suitable corrections to take into account the gaseous volume change that takes place as a result of reaction.

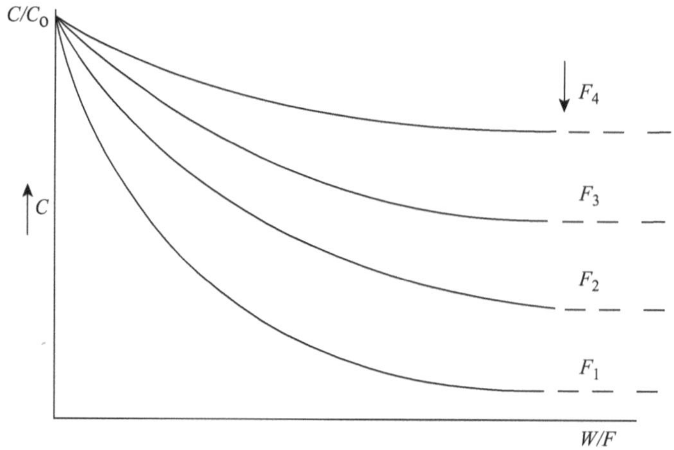

Figure 5.7:
Plots of C/Co vs. *W/F* for different flow rates (*F*) through a packed bed.

the so-called packing density of the catalyst and so it is more practical to consider the weight of the catalyst from the beginning of the bed to any position of the bed (W_X) and relate it to the total weight of catalyst used (W_T) and then assume a packing density. Although it is in principle possible to measure the concentrations of reactants and products throughout the bed (e.g. using a moveable probe or by providing sampling points along the length of the bed), this is difficult to achieve. In consequence, it is more common to measure the concentrations of gases at the exit of the bed as a function of the rate of flow through the catalyst[6] and then construct plots of the exit concentration as a function of *W/F* (see Figure 5.7). It is also possible to carry out a series of experiments each at different temperatures and to use these to give information on the activation energy for the reaction.

In principle, it is possible to fit the data corresponding to plots of the sort shown in Figure 5.7 to an integrated rate expression and hence to obtain values for the rate constant at various temperatures. In practice, this is difficult as the kinetic expressions are often not easily integrated (see Chapter 6). Hence, it is more common to use a "differential approach" by carrying out experiments to obtain the rates of the reaction under different conditions. The initial rate of the reaction shown in Figure 5.7 can be obtained by taking a tangent to the curve as *W/F* tends to zero, this requiring a separate set of experiments for each value of rate. However, if the measured conversion is low (less than about 20%), the conversion approximates to the rate at

[6] In principle, the weight of catalyst can also be varied. However, that is more difficult to achieve in a way that does not introduce errors. For example, it is difficult to make sure that the flow conditions remain the same for different bed packings and there is a danger of introducing channelling, particularly at low bed weights. Such problems can be largely avoided by using a catalyst bed that is diluted by different quantities of a suitable inert material, making sure that the total bed weight remains constant throughout the set of experiments.

the input conditions and so a single measurement suffices.[7] Hence, a set of rate data can be accumulated as a function of inlet concentrations and reaction temperature; further, by adding different amounts of product to the reaction mixture, the effects of the concentrations of the products can also be determined.

The preceding paragraphs assume that the catalyst particles in the packed bed being used are all easily accessible to the reactants and that reaction occurs homogeneously throughout each grain of the catalyst. In other words, there are no diffusion limitations. Such measurements are therefore restricted to experiments with relatively small particles (in which diffusion restrictions can be neglected) or to reactions with relatively low rates. In Chapter 7 we will discuss the use of larger catalyst geometries (e.g. pellets) of the type used in larger reactors and the problems introduced by mass transfer of reactants and products to and from the external surface of the catalyst sample (external diffusion) and also within the pores of catalyst pellets (internal diffusion) and we will see how these effects manifest themselves in kinetic measurements. We will now discuss several other types of reactors.

5.5 Fluidized Bed Reactors

Although fluidized bed reactors are generally used for large-scale applications (see Chapter 7), they can also be used for laboratory or bench-scale tests of a catalyst and so they will be discussed briefly in this chapter. If small catalyst particles are placed in a packed bed, severe pressure drop problems can be encountered, especially if higher gas flow rates are used. This problem can be circumvented by using a fluidized bed reactor. In such a system, the catalyst particles are placed in a vessel which has a porous plate at the bottom through which the reactant gases pass, as shown in Figure 5.8. At low gas velocities, gas bubbles will pass through the powdered catalyst bed in rather the same way that gas bubbles would pass through a liquid. However, as the gas velocity is increased, the bed becomes fluidized, increasing the volume considerably, and gas bubbles are no longer visible; in this situation, the catalyst bed is still well defined, with a separate gas phase above it. As long as the gas velocity is not increased too much, the particles remain in the reactor without any spillover to the exit tube.[8] All the many forms of fluidized bed reactors have a significant advantage over fixed bed reactors: that additional catalyst can be added to the reactor using a feed system at the top of the reactor and spent catalyst can be removed from the base of the reactor. The item on fluidized bed reactors in Wikipedia gives the advantages and disadvantages of such systems (http://en.wikipedia.org/wiki/Fluidized_bed_reactor) but it should be stressed that this item also covers the use of fluidized beds for other than catalytic reactions. The advantages of a fluidized bed system include the elimination of heat

[7] If the conversion is low, then the gradient of the chord joining the measured conversion to the origin approximates to the tangent at the origin.

[8] At even higher gas velocities, the gas and solid are not separated by gravity in the reactor but in a separate cyclone, as in the entrained reactor of the FCC process discussed in the next paragraph.

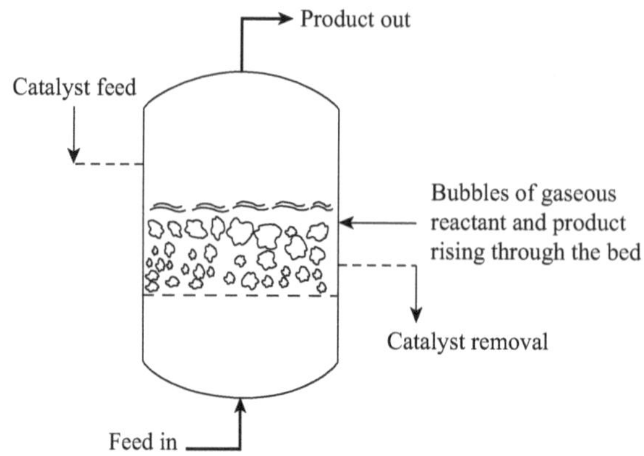

Figure 5.8:
Schematic representation of a fluidized bed reactor. The reactor can be operated with or without a through flow of catalyst; the latter mode is used if gradual deactivation of the catalyst occurs.

and mass transfer limitations while the disadvantages include the fact that the reactor is much larger than a packed bed reactor and involves higher energy costs and the possibility of particle entrainment. The following web page gives an animated representation of an operating fluidized bed: http://www.ansys.com/industries/chemical-process-fluidized-bed-ani.htm.

It should be stressed that fluidized bed reactors are more commonly used in industrial applications than in fundamental laboratory studies as they are suitable for practical large-scale operations. However, full-scale fluid bed reactors present difficulties in detailed modelling and so pilot-scale facilities are almost always used prior to introducing them in practice.

Figure 5.9 shows an entrained flow (or riser) reactor of the type used in the fluid catalytic cracking (FCC) process (see also Chapter 8). In this case, the gas flow rate is so high that the catalyst particles become entrained in the flow and pass through the reactor at the same rate as does the gas. On leaving the vertical (riser) part of the reactor, the gas and catalyst are separated in a cyclone; the gaseous product is removed from the system while the catalyst is recirculated to the reactor. If required, a separate regeneration volume can be incorporated (e.g. to burn off coke formed), spent catalyst can be removed from the system, and additional catalyst can be added. In some applications, additional components can be added to the catalyst stream; for example, an adsorbent can be added that will selectively remove a poison such as H_2S from the gas stream or an additional catalyst can be added to facilitate reactions parallel to the main catalytic process. One such example is the addition of a de-NOx components to help catalyse the selective reduction (in the riser section) of any NOx formed in the catalyst regenerator section. As with the more conventional fluidized-bed reactors discussed above, the use of a riser reactor is generally restricted to commercial operation, although smaller pilot plants are also used to help establish the operation of such systems.

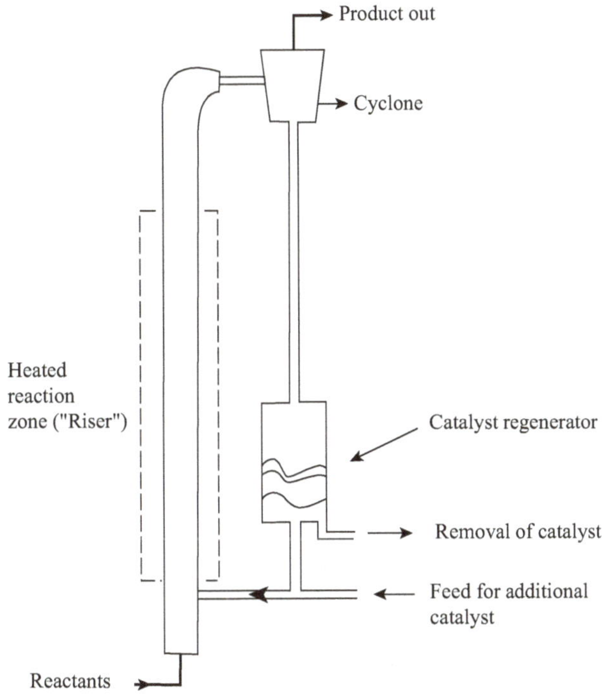

Figure 5.9:
Schematic representation of a riser reactor. The catalyst particles are entrained in the flowing reactant gas and separated after reaction in a cyclone. The catalyst particles are then regenerated in a separate regenerator section.

BOX 5.5 Recirculating Solids Reactor

The principle of the entrained reactor in which small particles are recirculated after passing through a regenerator has been extended by DuPont to the use of larger catalyst grains in a selective oxidative reaction, the oxidation of butane to maleic anhydride:

$$C_4H_{10} + {}^{7}/_{2}O_2 \rightarrow C_4H_2O_3 + 4H_2O$$

In this system, the oxidation step takes place in one reactor using the oxygen of the catalyst; the partially reduced catalyst is then transported to a second reactor in which it is re-oxidized using air. The advantage of the system is that the maleic anhydride is not exposed to oxygen and the production of undesired by-products such as CO_2 is minimized. The process requires mechanical stable catalysts which are not abraded during the transportation process. The energy costs involved in the transportation of large quantities of catalyst are also very high (see Task 5.2).

5.6 Pulse Reactors

A pulse reactor is a system in which a small pulse of the reactant or reactants of interest are passed over a small sample of the catalyst being examined and then directly into the analysis system.[9] A typical set-up is shown in Figure 5.10; in this, the analysis is carried out using a gas chromatograph in which the column separates the mixture leaving the reactor into its separate components and the detector, after suitable calibration, gives the appropriate concentrations. Such systems have generally been used for rapid screening of a series of catalysts to find the optimum for a given reaction. However, what is possibly a more valuable application is its use in tracer experiments in which a labelled molecule is added to the reaction mixture to see where the label turns up in the products of the reaction. The technique was pioneered by Emmett and his colleagues[10] in their work on the mechanism of the Fischer Tropsch synthesis of hydrocarbons from syngas ($CO + H_2$ mixtures). The use of such systems has been reviewed by Galeski and Hightower.[11] The method has the drawback that the catalyst is not examined under steady-state conditions and so any conclusions reached apply only to the

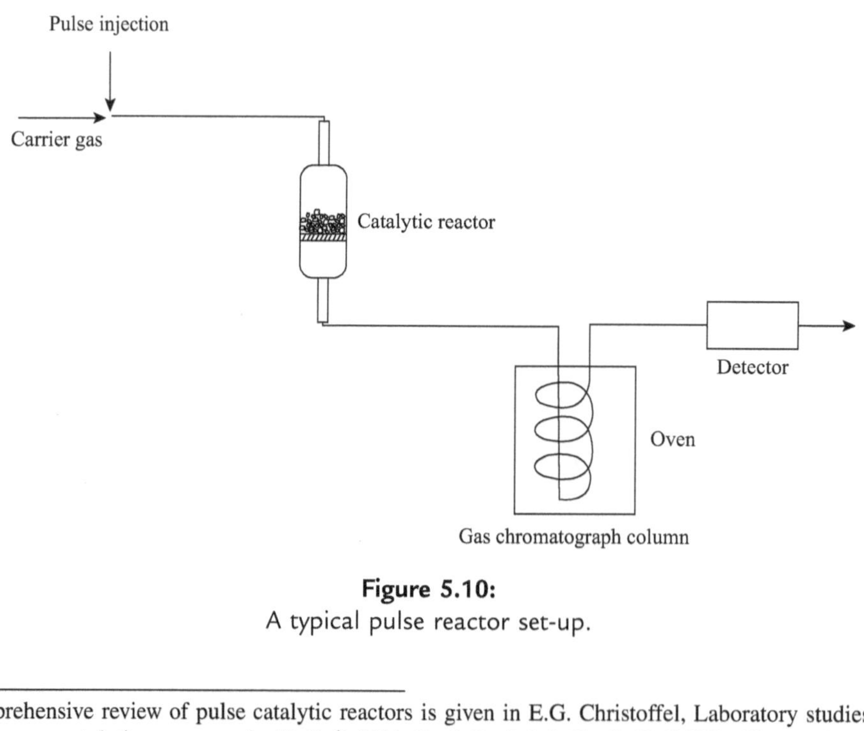

Figure 5.10:
A typical pulse reactor set-up.

[9] A comprehensive review of pulse catalytic reactors is given in E.G. Christoffel, Laboratory studies of heterogenous catalytic processes, in: Z. Paál (Ed.), Stud. Surf. Sci. Catal. 42 (1989), Chapter 4.

[10] R.J. Kokes, H. Tobin, P.H. Emmett, New microcatalytic-chromatographic technique for studying catalytic reactions, J. Am. Chem. Soc. 77 (1955) 5860.

[11] J.B. Galeski, J.W. Hightower, Microcatalytic reactors: Kinetics and mechanisms with isotopic tracers, Canadian J. Chem. Eng. 48 (1970) 151.

catalyst under the non-steady-state conditions used; in other words, the data often refer to a catalyst which has not been fully activated or for which the surface conditions change during the passage of the pulse. The technique is seldom used nowadays although its equivalent is sometimes used in combinatorial and high-throughput studies in which the activities of multiple small samples of potential catalysts are compared by exposing each briefly to pulses or flows of reactants (see Box 5.6).

As with pulse methods, high-throughput methods give only approximate data since the conditions used to monitor the activities and selectivities of the catalysts produced are far from ideal. Further, the approach, although saving considerable manpower, is dependent on good data-handling methods being applied. Some of the papers from the EuroCombiCat meetings (Box 5.6) are worth examining to get a feel for the level of sophistication of the methods. One of the problems of assessing the effectiveness of the methods is that although many claims have been made that new commercial catalysts have been developed as a result of work using the methods, most of the critical results are shrouded in industrial secrecy. An example of a company making systems for high-throughput experimentation is HTE based in Heidelberg, a subsidiary of BASF. (See http://www.hte-company.com/en/sitemap for further information

TASK 5.2 DuPont's Circulating Fluidized Bed Technology

Examine some of the literature on this process, starting with the work of R.M. Contractor (Dupont's CFB technology for maleic anhydride, Chem. Eng. Sci. 54 (1999) 5627).

BOX 5.6 Combinatorial and High-Throughput Catalysis

The search for new catalysts for specific reactions can be time consuming and expensive in manpower as well as in the use of equipment and chemicals if multiple flow reactors are used to measure reaction kinetics. Hence, much effort has been devoted to the use of the so-called combinatorial and high-throughput techniques to examine multiple samples of catalysts prepared in such a way as to provide a statistical survey of potentially effective catalysts. Many different types of approaches have been used, from making libraries of small samples and testing them in a series of parallel reactors simulating more conventional plug-flow reactors (see, e.g., G. Grasso, J.R.H. Ross, in: J.R. Sowa (Ed.), Catalysis of Organic Reactions, Taylor and Francis (Baton Rouge), 2005, 393) to making catalyst samples as multiple spots on a plate and then monitoring the activity of each sample in some way, for example, using pulses of reactant directed to each sample in turn and monitoring the resultant products with a snorkel device. There is a series of scientific meetings on the subject, EuroCombiCat (http://www.eurocombicat.org and the proceedings of two of the meetings are to be found in Catalysis Today, 137 (2008), Issue 1 and 159 (2011) Issue 1.

on the systems available and some technical data sheets.) An early company concentrating on this technology was Symyx Inc. of Santa Clara, California, which has recently merged with Accelrys (http://accelrys.com).

TASK 5.3 High-Throughput Experimentation

One of the founders of high-throughput experimentation was W.H. Weinberg of Symyx. Using his name, search Web of Science or Scopus for some of his papers and use the citations to these in order to locate other groups who have published in the area. Examine some of the types of approaches used and catalytic systems studied. Wherever appropriate, use also the references in Box 5.6.

5.7 The TAP Reactor

The technique known as temporal analysis of products (TAP) was developed by John Gleaves and his team in Monsanto in 1978–1979 in order to speed up their search for new catalysts for specific processes. It was introduced to the scientific community in 1984 and its use was covered by a number of patents. In essence, it is a method that depends on the pulsing of very small amounts of the reactant gas or gases, either simultaneously or after defined time lapses, onto the surface of a catalyst using high-speed pulse valves (Figures 5.11 and 5.12) and measuring the responses using a quadrupole mass spectrometer (QMS), which is located very close to the exit of the catalyst bed. The system can operate in a number of different modes, depending on whether a small number of pulses are used ("probe experiments", providing intrinsic kinetic information) or larger numbers of pulses

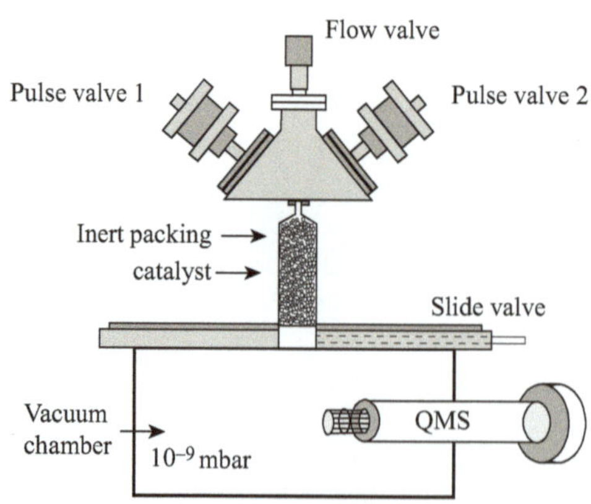

Figure 5.11:
The TAP 2 reactor. *Source: J. Pérez-Ramírez and E.V. Kondratenko, Catal. Today, 121 (2007) 160–169. Reproduced with kind permission of Elsevier.*

Figure 5.12:
Photograph of the TAP 2 reactor installed in ICIQ, Tarragona, and Spain (manufactured by Autoclave Engineers.) *Source: J. Pérez-Ramírez and E.V. Kondratenko Catal. Today 121 (2007) 160. Reproduced with kind permission of Elsevier.*

are supplied in sequence ("multipulse experiments", giving information on adsorption and dissociation of different species, or "pump-probe experiments", involving admission of two or more reactants at different time intervals and giving information on the life-time, and reactivity and selectivity of adsorbed species). The equipment used is very expensive, depending on stainless steel ultrahigh vacuum systems (Figure 5.12) and a very limited number of systems have been built to date. Nevertheless, the experiments carried out in these systems has given valuable information (often confirming earlier ideas based on less definitive experiments) on a wide range of reaction systems.[12]

TASK 5.4 The TAP Reactor

Use Web of Science or Scopus to identify the papers by J.T. Gleaves (Washington University). Identify the main authors working in the field. Then, use the listing papers that report the use of TAP to study the selective oxidation of butane to maleic anhydride. Summarize the main conclusions of these papers in relation to the probable mechanism of the reaction.

[12] A special issue of Catalysis Today (vol. 121, issues 3–4 (2007) edited by E.V. Kondratenko and J. Perez-Ramirez), includes a very useful summary of the uses of the technique and contributions from some of the main proponents.

5.8 SSITKA

The technique known as SSITKA (steady-state isotopic transient kinetic analysis), developed originally by Happel, Bennet and Biloen,[13] involves switching the isotopic composition of the gas flowing over a catalyst in a packed bed without any (significant) all-over change in the gas-phase composition. This is done in a system in which it is possible to switch from one isotope under investigation to the other while not disturbing the other flows. Figure 5.13 shows a typical experimental set-up for an investigation of the oxidation of methane over a Pd/alumina catalyst. A flow containing $^{16}O_2$ with a trace of Ar as an inert tracer is displaced by a flow of $^{18}O_2$ and the change in the isotopic distribution of the product CO_2 is measured relative to the disappearance of the signal due to Ar. Typical results obtained at a temperature of 583 K are shown in Figure 5.14. From these and other results, it was possible to conclude that the oxygen of the catalyst was used in the reaction and that a Mars-van Krevelen mechanism applied to the reaction. Full treatment of the data of SSITKA experiments requires very careful mathematical analysis. Interested readers should examine the literature on the subject, using for example, one of the references given in footnote[13] to trace some of the work on the subject. Further treatment is beyond the scope of this textbook.

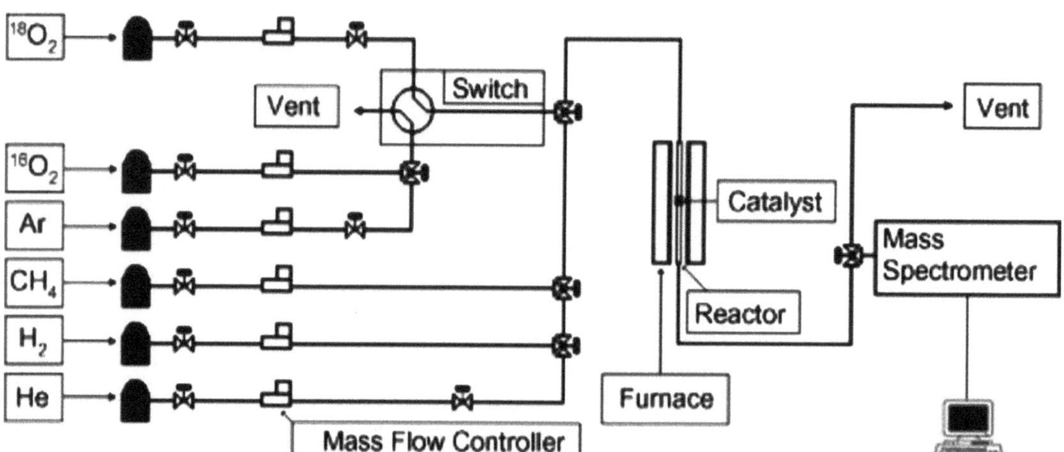

Figure 5.13:
An apparatus used for SSITKA measurements on the catalytic oxidation of methane over a Pd catalyst. *Source: A. Machocki, M. Rotko, B. Stasinska, Catal. Today 137 (2007) 312. Reproduced with kind permission of Elsevier.*

[13] J. Happel, Chem. Eng. Sci. 33 (1978), 1567; C.O. Bennett, ACS Symp. Ser. 178 (1982) 1; J. Biloen, J. Molec. Catal. 21 (1983) 17.

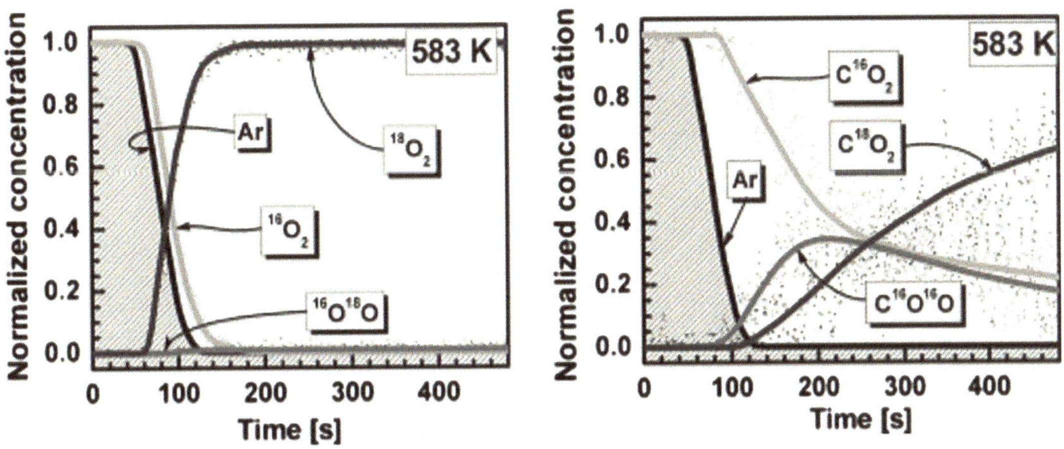

Figure 5.14:
The results of switching from $16O_2$/Ar/CH_4/He to $18O_2$/CH_4/He at the temperature of 583 K, ensuring 10% of methane conversion. *Source: Reproduced from A. Machocki, M. Rotko, B. Stasinska, Catal. Today 137 (2007) 312. Reproduced with kind permission of Elsevier.*

5.9 "In Situ/Operando" Methods

The SSITKA technique described in the previous section is aimed at gaining information on the mechanism of a catalytic reaction under operating conditions and generally depends on curve-fitting but does not involve any direct observation inside the reactor. "In situ" and the so-called Operando methods aim to identify reacting species directly under catalytic conditions.[14] A number of physical identification techniques have been used in conjunction with such measurements, these including transmission Fourier transformation infrared spectroscopy (FTIR), diffuse reflectance infrared Fourier transform spectroscopy (DRIFTS), Raman spectroscopy, UV-visible spectroscopy, X-ray photo-electron spectroscopy (XPS), sum-frequency generation (SFG), vibrational spectrosopy and NMR. Although all of these methods are capable of monitoring surface species directly while carrying out a catalytic experiment, the most commonly applied method involves DRIFTS.

As an example of the application of operando spectroscopy, we will take the case of the use of Raman spectroscopy[15] together with on-line gas chromatographic analysis of the gas phase to study the oxidative dehydrogenation of propane over VO_x/Al_2O_3 catalysts. The equipment

[14] The term "in situ" implies that the measurements have been carried out using an operating catalyst but no attempt has been made to relate the measurements to actual catalytic measurements; on the other hand, "Operando" measurements are those in which there is on-line analysis of reaction effluent to ensure that the activity data obtained in the operando reactor are consistent with those observed in a conventional reactor.

[15] Raman spectroscopy on its own provides information on the nature of oxide surface species; in this case, it is able to distinguish clearly between different forms of V–O bonds on the surface.

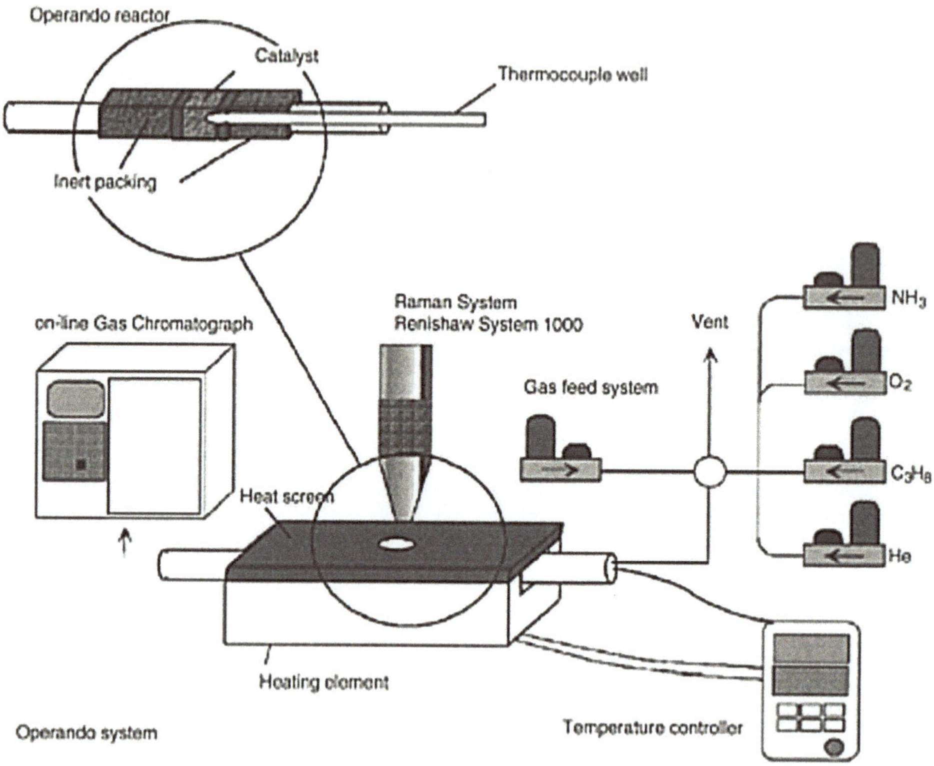

Figure 5.15:
An operando Raman-GC set-up. *Source: M.O. Guerrero-Peŕez, M.A. Bañares, Catal. Today 113 (2006) 48–57. Reproduced with kind permission of Elsevier.*

used by Guerrero-Pérez and Bañares is illustrated schematically in Figure 5.15 and the results of experiments carried out at different propane to oxygen ratios are shown in Figure 5.16. The only surface species visible in all the different gas compositions are V = O, although the band at 1009 cm^{-1} corresponding to isolated V = O bonds does disappear at a propane/oxygen ratio of 2.0. Parallel work using UV–vis spectroscopy showed that there was significant reduction of the catalyst during a change of C_3H_8/O_2 ratio from 10/1 to 2/1. Hence, the authors conclude that the alteration is due to reduction of surface polymeric vanadium oxide species and that only the V = O surface species are active in the oxidative dehydrogenation reaction.

TASK 5.5 Operando Spectroscopy

Using Scopus or Web of Science, examine the papers included in the two conference proceedings mentioned in Box 5.7 and, for the most significant ones, tabulate the authors, the reactions studied, the analytical methods applied and the main conclusions reached.

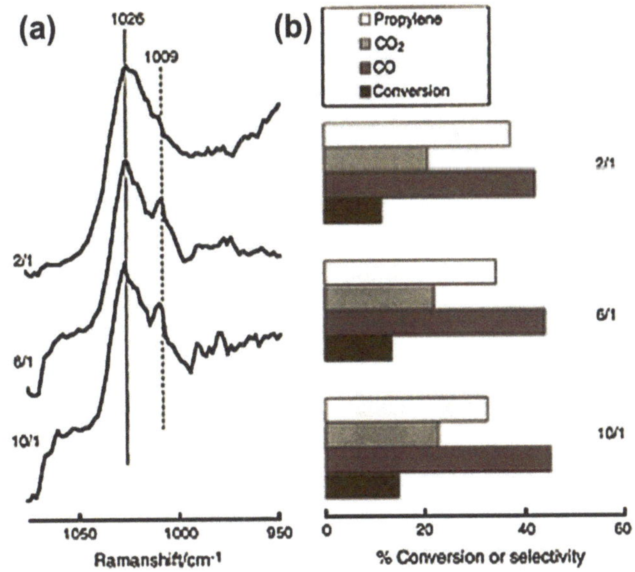

Figure 5.16:
(a) Operando Raman spectra of a VO_x/Al_2O_3 catalyst during oxidative dehydrogenation of propane at 400 °C as a function of C_3H_8/O_2 molar ratio. (b) Corresponding conversions of C_3H_8 and selectivities to CO, CO_2 and C_3H_6. *Source: M.O. Guerrero-Pérez, M.A. Bañares, Catal. Today 113 (2006) 48–57. Reproduced with kind permission of Elsevier.*

BOX 5.7 Some Conferences on Operando Spectroscopy

A series of International Congresses on Operando Spectroscopy have been held, the first in Lunteren, the Netherlands, in 2003; the fourth in the series is to be held in Brookhaven National Laboratory in 2012 (http://www.nsls.bnl.gov/newsroom/events/workshops/2012/OperandoIIV/default.asp). Two issues of Catalysis Today have been devoted to meetings on the subject: Vol. 113 (1–2) (2006), Guest Editors, F. Meunier and M. Daturi, being the proceedings of the Second International Congress (Operando 2) held in Toledo, Spain, in 2006. (The third, based on Operando 3, held in Rostok, Germany, in 2009, is in preparation). These two issues contain papers based on the use of each of the analytical techniques mentioned above and give good examples of their application to the study of a variety of reactions.

5.10 Microreactor Methods

There have been very significant developments in the application of microreactor techniques and applications over the last two decades. The development of microchannel reactors has opened up new possibilities for the design of small and efficient chemical plants in which catalysis plays a pivotal role. Such developments have depended on improved engineering

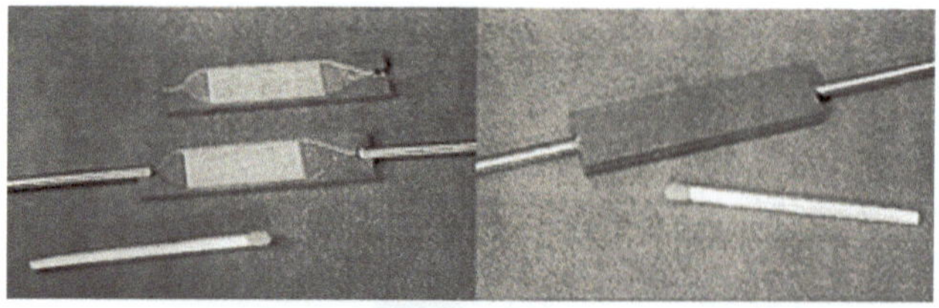

Figure 5.17:
Microchannel Reactor Platelets used for the Catalytic Combustion of Methane. *Source: M. O'Connell, G. Kolb, R. Zapf, Y. Men, V. Hessell, Catal. Today 144 (2009) 306. Reproduced with kind permission of Elsevier.*

techniques that allow the creation of channels with very small dimensions on the surfaces of metallic plates, these plates then being stacked one on top of another to create a miniaturized reactor system. Techniques have been developed to prepare catalysts within the channels in such a way that the materials have very similar properties to those of conventional catalysts. Such systems have a number of significant advantages over more conventional reactors, these including improved heat and mass transfer and ease of scale-up by simply increasing the number of plates in the assemblies. Further, because of their small volumes, they can safely be used in situations in which explosive situations might be encountered.

Many potential applications of microreactors have been reported in the literature. One of the most promising developments is in the area of the production of hydrogen for mobile fuel cell applications. This is achieved by carrying out steam reforming of methane or autothermal reforming of higher hydrocarbons (diesel or petroleum). These processes, which will be discussed in more detail in Chapter 8, are endothermic and so heat has to be supplied to the system. Work is therefore being carried out to find systems to bring about the catalytic combustion of the anode off-gases from a fuel cell (containing hydrogen and methane) to provide the necessary heat. Figure 5.17 shows the construction of microchannel platelets, which have been developed for this purpose. These were coated with alumina (wash-coated into the channels) and then Pt/Mo or Pt/W active components were introduced. The system was then tested for the combustion of methane; several commercial catalysts were also examined. The results of such tests are shown in Figure 5.18. For the purposes of this chapter, all that need to be noted from these results is that all the catalysts give 100% conversion of methane at temperatures above about 600 °C and that the selectivity to CO_2 under these conditions is 100%.

TASK 5.6 Microreactor Technology

One of the pioneers in this field has been V. Hessel, formerly of the Institute of Microreactors in Mainz, Germany and now a full-time professor at the Technical University of Eindhoven, The Netherlands. Carry out a Scopus or Web of Science search on his papers, then identify other authors working in the field and learn something of the types or reactor and reactions studied.

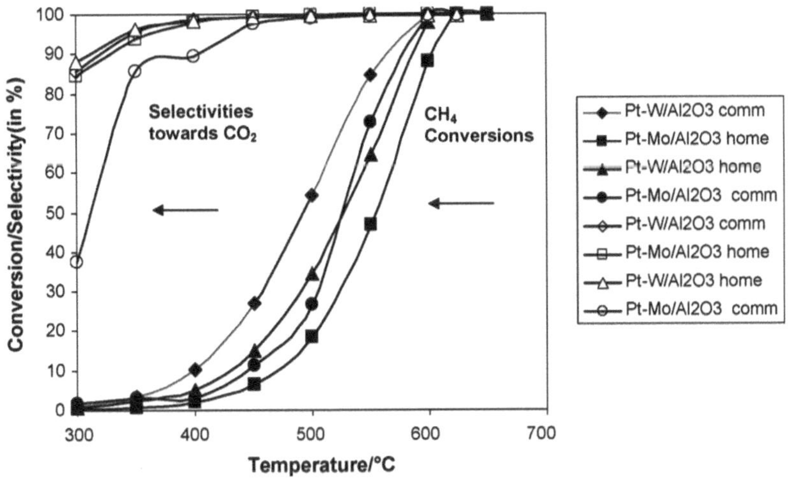

Figure 5.18:
Methane conversions and selectivities for four catalysts. For conditions, see source article: *M. O'Connell et al., Catal. Today 144 (2009) 306–311. Reproduced with kind permission of Elsevier.*

5.11 Conclusions

The majority of the types of reactor discussed in this chapter are of the sort that are used to examine the catalytic behaviour of catalysts on a laboratory scale, either to allow comparison between different catalysts as part of a development programme or to give information on the kinetics and mechanism of the reaction. The kinetics and mechanisms of catalytic reactions will be discussed in Chapter 6. In Chapter 7 we will discuss other aspects of the use of catalytic reactors: their use in full-scale processes and the problems associated with heat and mass transfer of the reactants.

CHAPTER 6

The Kinetics and Mechanisms of Catalytic Reactions

Chapter Outline

6.1 Introduction

Kinetics and mechanism are closely interrelated and kinetic measurements are frequently used to give information on the mechanisms of reactions. Additionally, detailed kinetic equations are important in relation to the design and operation of chemical reactors. The approach is generally quite straightforward for homogeneous systems but the situation is very frequently more complex in the case of the kinetics of catalytic reactions. This is because the expressions which are determined experimentally can be much more complex than those usually associated with simple homogeneous reactions.[1] A consequence, particularly for engineering applications, is that simplified kinetic expressions ("intrinsic kinetics") are used which describe the operation of the catalyst very adequately over restricted ranges of the partial pressures of the reactants and products and also of temperature but which bear little formal relationship to the mechanism of the reaction in question. The situation is even more complex if physical phenomena such as mass and heat transfer play a part in determining the observed rates. A discussion of the latter phenomena is included in Chapter 7.

[1] There are exceptions to every rule. For example, the formal kinetic expressions for chain reactions are just as complicated as those for catalytic reactions.

Heterogeneous Catalysis. DOI: 10.1016/B978-0-444-53363-0.10006-4

In Chapter 3, we discussed how a catalyst works: by adsorbing the reactants on sites on its surface and by providing an alternative route by which the reaction can occur that is more energetically favourable compared to that taken in an equivalent homogeneous process. Having then discussed in Chapters 4 and 5 how to make a catalyst and to characterize it as well as to measure its catalytic properties, we examine in this chapter how the kinetics of a catalytic reaction may be described mathematically, both mechanistically and more globally. We then proceed to discuss the mechanistic aspects of some reactions that have been studied in detail.

We will start by considering in the following sections three commonly encountered formal kinetic expressions that describe the kinetics of simple model reactions in a way that relates to the suggested mechanisms ("mechanistic kinetics"). All of these approaches use the concepts we have encountered in relation to the Langmuir isotherm describing the adsorption of the reactants on the surface of a uniform catalyst surface and extend these concepts to explain the rates of reaction of these adsorbed reactants. We will then in subsequent sections discuss some application of these mechanisms to real catalytic systems and will give suggestions for further reading on these and other reactions.

6.2 Unimolecular Reaction of Reactant A to Give Products

In Chapter 2, we described the derivation of the Langmuir adsorption isotherm (Eq. 2.8) that relates the quantity of gas A adsorbed on a uniform surface at a constant temperature to the pressure of the gas P_A above the surface:

$$V_{ads}/V_m = \theta_A = b_A P_A/(1 + b_A P_A) \tag{6.1}$$

where V_m is the monolayer capacity of the surface. The coverage θ_A is a measure of the concentration of A on the surface.

Let us now consider the rate of a reaction:

$$A_{(g)} \rightarrow P_{(g)} \tag{6.2}$$

occurring over such a uniform surface, where the subscript (g) indicates that the species in question is in the gas phase. We then set up a scheme in which A becomes adsorbed on the surface as A_{ads}, where the subscript ads indicates that the species is adsorbed. We then consider that the rate of reaction of the surface species A_{ads} to give the gaseous product $P_{(g)}$ is rate determining and that this step is much slower than the adsorption and desorption steps so that we can consider the first step as being effectively at equilibrium:

$$A_{(g)} \rightleftarrows A_{ads} \rightarrow P_{(g)} \tag{6.3}$$

We can then write that the all-over rate, r_A, of the reaction is proportional to the concentration of A_{ads}, this being given by θ_A, determined by the appropriate Langmuir isotherm:

$$r_A = -dc_A/dt = k_T \cdot \theta_A = k_T b_A P_A/(1 + b_A P_A) \tag{6.4}$$

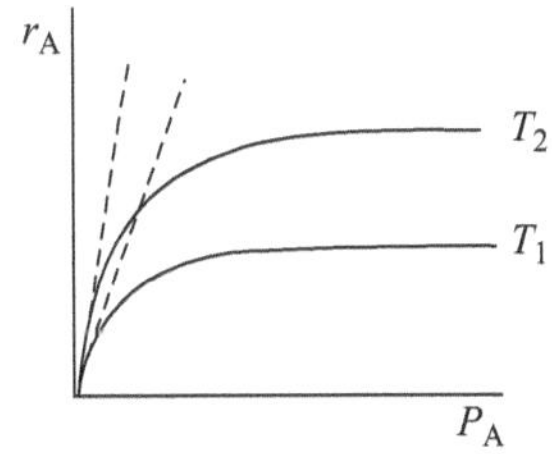

Figure 6.1:
Plot of the rate of reaction 2.3 vs. P_A for two temperatures, T_1 and T_2.

where r_A is the rate of disappearance of A, k_T is the rate constant at any temperature, and b_A is the adsorption constant for the adsorption of A; b_A is essentially the equilibrium constant for the first step in Eq. (6.4). Figure 6.1 shows a plot of this simple expression in the form of r_A versus the pressure of A, P_A, for two different temperatures, T_1 and T_2 ($T_2 > T_1$). At low pressures, each curve approximates to a straight line with gradient $k_T\, b_A$ while at high pressures each curve tends to being a straight line parallel to the P_A axis. In other words, the low pressure data correspond to a first order equation:

$$r_A = k_T \cdot b_A P_A \tag{6.5}$$

while the high pressure data approximate to a zero order equation:

$$r_A = k_T \tag{6.6}$$

In between the two extremes, the full expression (6.4) applies. As we shall see later, it is possible to approximate to this expression, over very limited ranges of pressure, by an expression:

$$r_A = k_{T'} P_A^n \tag{6.7}$$

where n is a number between 0 and 1.[2]

Figure 6.2 shows a so-called reciprocal plot of Eq. (6.4), in other words a plot of $1/r_A$ vs. $1/P_A$, for one of the sets of data of Figure 6.1. As shown, the values of k_T and b_A can be obtained from the intercepts on the two axes (or from one of the intercepts and the gradient). Any deviation from the straight line found would arise from deviations from the ideal behaviour assumed in setting up the equation, for example due to a variation in the strength of the bond between A and the surface when the coverage θ_A changes with pressure.[3] By obtaining similar plots at

[2] This is equivalent to making use of the Freundlich isotherm (Eq. 2.3) instead of the Langmuir isotherm to describe the adsorption equilibrium.

[3] The approach we have used here makes the simplifying assumption that the adsorbed species are in equilibrium with the gas phase and that the rates of adsorption and desorption are much faster than the rate of the reaction of the surface species to give the gaseous product, $P_{(g)}$. An alternative more rigorous approach to setting up the rate expression uses the method of the stationary state to give an equivalent expression that includes both the rate constants for adsorption and desorption as well as that for the formation of $P_{(g)}$. Then, as long as the magnitude of the rate constant for the catalytic reaction is small compared to those for adsorption and desorption, the two approaches give the same result.

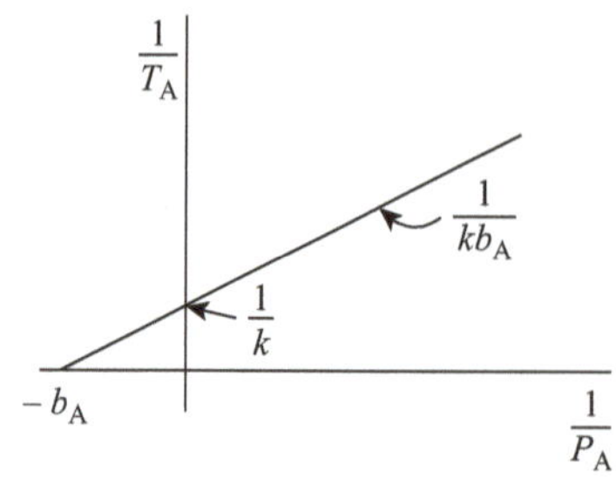

Figure 6.2:
Plot of rate vs. $1/T$ for the data of Figure 6.1 at one temperature.

a variety of temperature, values for both the activation energy for the reaction, E_A, and the heat of adsorption of the reactant A, ΔH^o_{ads}, can be obtained (from plots of the values of $\ln k_T$ and $\ln b_A$ vs. $1/T$ respectively).[4]

TASK 6.1

What would be the position of the reciprocal plot in Figure 6.2 for the second curve of Figure 6.1, assuming that the one shown is for the data for a temperature T_1.

BOX 6.1 Enzyme Kinetics

An enzyme is also a catalyst and it contains active sites, albeit much more complex in structure, equivalent to those of a heterogeneous catalyst. An enzyme-catalysed reaction is frequently depicted as occurring by a sequence of steps involving the enzyme E, the substrate S (i.e. the reactant) and the product P:

- the formation of an enzyme-substrate complex:

$$E + S \rightarrow ES \qquad \text{rate constant } k_a$$

- the reverse reaction:

$$ES \rightarrow E + S \qquad \text{rate constant } k_a$$

- the unimolecular formation of the products and release of the enzyme:

$$ES \rightarrow P + E \qquad \text{rate constant } k_b$$

[4] If only two values of k_T and b_A are known, the values of E_A and ΔH^o can be obtained by solving two simultaneous equations. For example, the value of E_A is given by the equation $E_A = \{2.303R\ T_1T_2/(T_2 - T_1)\}\{\log_{10}k_2 - \log_{10}k_1\}$, where R is the gas constant.

BOX 6.1 Enzyme Kinetics—*Cont'd*

The kinetics of an enzyme-catalysed reaction is conventionally described by the so-called Michaelis—Menton equation:

$$\text{Rate of formation of } P = k_b[S][E_o]/([S] + K_M) \tag{1}$$

where K_M, the Michaelis constant, is given by

$$K_M = (k'_a + k_b)/k_a \tag{2}$$

and $[E_o]$ is the total concentration of enzyme, both bound and unbound.

This equation was derived by Michaelis and Menton in 1913 assuming a rapid equilibrium in the way that we have done above for a heterogeneous catalyst; in the case of a rapid equilibrium, $k'_a >> k_b$ and hence $K_M = k'_a/k_a$. K_M is therefore equivalent in all ways to b_a, the adsorption constant in the Langmuir approach.

It is interesting to note that the method of linearizing the Michaelis—Menton equation leading to plots of the type shown in Figure 6.2 is ascribed to Lineweaver and Burk and that these are known as Lineweaver—Burk plots. Many books have been devoted to the determination of enzyme kinetics and to the effects of inhibitors, etc. An interesting example of animated diagrams related to enzyme kinetics is found at http://www.wiley.com/college/pratt/0471393878/student/animations/enzyme_kinetics/index.html\ These could equally well apply to a heterogeneously catalysed reaction of the type discussed above.

It is also interesting to note that J.J. Carberry has pointed out in his book entitled *Chemical and Catalytic Reaction Engineering* (reprinted by McGraw Hill, NY (1976), partially available at http://books.google.com/books?id=arJLaKa4yDQC&source/), that the derivation of the rate expression for the case of enzyme kinetics described above was first carried out by V.C.R. Henri in 1902 (Acad. Sci. Paris, 135 (1902) 916); he also pointed out that Henri derived the Langmuir isotherm and the resulting rate equation. It is often complained that the current generation of young researchers is unfamiliar with the earlier literature and this is blamed on the use of electronic searches which are based on limited data. The example treated above is an excellent example of how original references can become obscured by time and over-familiarity with a specific topic, even in a time period when electronic methods were unheard of. The whole subject of communication between scientists in the early twentieth century is one which deserves much more attention.

We have assumed for the unimolecular reaction (Eq. 6.4) that only the molecule A is adsorbed on the surface. However, it is possible that the product P is adsorbed in significant amounts before desorbing to the gas phase and that there is an equilibrium for the desorption step equivalent to that for the adsorption step:

$$A_{(g)} \rightleftarrows A_{ads} \rightarrow P_{ads} \rightleftarrows P_{(g)} \tag{6.8}$$

The rate expression for the reaction is now modified by the inclusion of a term for adsorbed P in the denominator:

$$r_A = -dc_A/dt = k_T \cdot \theta_A = k_T \cdot b_A P_A/(1 + b_A P_A + b_p P_p) \tag{6.9}$$

where P_P is the partial pressure of the product gas in the reaction mixture. We have now encountered a situation never found in simple homogeneous elementary reactions in which the concentration of the product can have an effect on the rate of the reaction. If the adsorption of P is relatively strong when compared with that of A and $b_P P_P >> (1 + b_A P_A)$, Eq. (6.9) tends in the limit to:

$$r_A = -dc_A/dt = k_T \cdot \theta_A = k_T \cdot b_A P_A / b_p P_p \tag{6.10}$$

In other words, the reaction is now first order in A but it has an order of -1 in P. The product P is thus now effectively a "poison" for the reaction.[5,6]

6.3 Bimolecular Reactions – Langmuir–Hinshelwood Kinetics

Sir Cyril Hinshelwood (http://nobelprize.org/nobel_prizes/chemistry/laureates/1956/hinshelwood-bio.html/) was a British chemist best known for his work on the kinetics of chain reactions. In his book entitled *The Kinetics of Chemical Change*, first published in 1926, Hinshelwood expanded an earlier idea of Langmuir from 1921 to come up with what is now known as Langmuir–Hinshelwood kinetics[7] to explain the reaction between two molecules over the surface of a heterogeneous catalyst. Let us consider first the catalysed reaction:

$$A_{(g)} + B_{(g)} \rightarrow C_{(g)} \tag{6.11}$$

The coverage of the surface sites by A_{ads} and B_{ads} are given by the isotherms:

$$\theta_A = b_A P_A / (1 + b_A P_A + b_B P_B) \tag{6.12}$$

and

$$\theta_B = b_B P_B / (1 + b_A P_A + b_B P_B) \tag{6.13}$$

Assuming that the rate determining step for the reaction between $A_{(g)}$ and $B_{(g)}$ is given by the rate of the surface reaction of the two species A_{ads} and B_{ads} whose concentrations can be expressed by θ_A and θ_B, the rate of reaction can be written as:

$$r_A = -dc_A/dt = k_T \cdot \theta_A \cdot \theta_B = k_T \cdot b_A P_A \cdot b_B P_B / (1 + b_A P_A + b_B P_B)^2 \tag{6.14}$$

---

5 In enzyme kinetics, this would be referred to as "inhibition by products".

6 Poisons can also be added separately but their effects on the kinetics are exactly the same, there being a term $b_X P_X$ in the denominator for each poison species X. If X is very strongly adsorbed, the reaction can be totally inhibited.

7 In some American literature, the names Hougen and Watson are often also encountered associated with this approach (O.A. Hougen, K.M. Watson, Chemical Process Principles, vol. III, Wiley, New York (1943); the treatment is outlined in detail by Carberry (Chemical and Catalytic Reaction Engineering, Clarendon, Oxford (1940); reprinted by Wiley, NY (1976); see Box 6.1). According to Carberry, the approach, although similar in form to the Langmuir–Hinshelwood theory, is more explicit in terms of its inclusion of the effects of catalyst sites, of diffusion and of catalyst deactivation. We will, however, restrict ourselves here to the Langmuir–Hinshelwood approach as it is perfectly adequate for the present purposes.

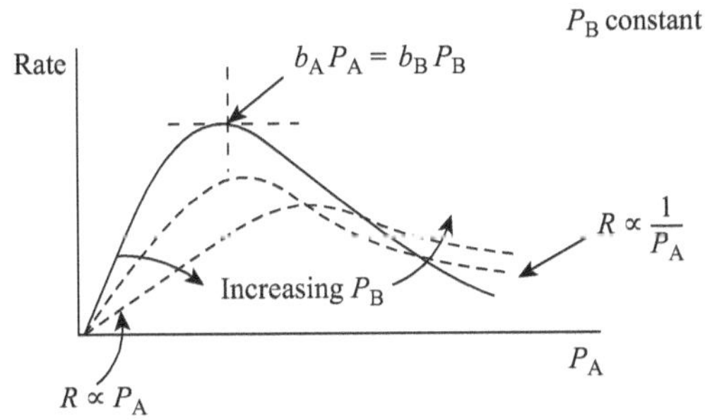

Figure 6.3:
Plot of the Langmuir–Hinshelwood model for a bimolecular reaction of A and B over a catalyst as a function of P_A with P_B held constant, showing the effect of increasing P_B.

Figure 6.3 shows a plot of the rate of the hypothetical reaction (6.10) as a function of the pressure of A, P_A, the value of the pressure of B, P_B, being kept constant. In the case of this biomolecular reaction, the rate goes through a maximum. At this point, $\theta_A = \theta_B$ and so the ratio of b_A to b_B can be established (since $b_A P_A = b_B P_B$ at that point). The figure also shows the effect of using larger values of P_B: the curves move to the right with a decrease in rate at lower values of P_A and an increase at higher values of P_A. At lower values of P_A, the rate is proportional to P_A while at higher values, the rate is inversely proportional to P_B.

If neither A nor B is strongly adsorbed, $1 >> b_A P_A + b_B P_B$ and so Eq. (6.14) reduces to:

$$r_A = -dc_A/dt = +dC_C/dt = k_T \cdot b_A P_A \cdot b_B P_B = k_T P_A \cdot P_B \tag{6.15}$$

In other words, the rate equation reduces to a second order rate expression.

When one of the reactants (e.g. A) is strongly adsorbed and the other (B) is weakly adsorbed, then the equation reduces to:

$$r_A = -dc_A/dt = +\ dC_C/dt = k_T \cdot b_A P_A / b_B P_B = k_T'' P_A / P_B \tag{6.16}$$

The situation is even more complicated if the products of the reaction are adsorbed to an appreciable extent or if there is a poison present in the reaction mixture. However, we will not consider such cases here as they result in even more complex expressions that are difficult to analyse in a satisfactory manner. Indeed, it is often possible to come up with several different mechanistic explanations for a complex set of kinetic data and it is only with additional supporting evidence of the sort provided by some of the methods described in Chapter 5 is it possible to decide on the most likely mechanism. Before moving on to discuss what happens when one or other, or both, of the reactants dissociates on the surface before reaction, we will first discuss two other kinetic expressions which have some importance in relation to catalytic mechanisms.

BOX 6.2 Effect of Temperature on the Rate of a Catalytic Reaction

For a homogeneous (non-catalysed) first order gas-phase reaction for which $-dC_A/dt = kC_A$, an increase in temperature only has an effect on the value of the rate constant k and application of the Arrhenius equation gives rise directly to the activation energy for the rate-determining step in the reaction. However, when we are dealing with a heterogeneously catalysed reaction such as one of those shown in Eqs. (6.3) and (6.8), the corresponding rate equations contain, in addition to rate constants, adsorption constants which also have temperature dependencies. Let us consider the case of Eq. (6.4):

$$r_A = -dc_A/dt = k_T \cdot \theta_A = k_T \cdot b_A P_A/(1 + b_A P_A) \quad (6.4)$$

At low values of P_A, the effective rate constant is $k_T\, b_A$ while at high values of P_A it is k_T. As b_A is effectively an equilibrium constant with an enthalpy change of adsorption of ΔH_a^o, its temperature dependence is given by the van't Hoff isochore:

$$d(\ln b_A)/dt = \Delta H_a^o/RT^2$$

or, in its integrated form:

$$b_A = C_{exp}(-\Delta H_a^o/RT^2)$$

Hence, at low P_A, the apparent activation energy E_{app} for the reaction is given by $E_a + \Delta H_a^o$ while at higher values of P_A, the apparent activation energy is the same as E_a, the "true" activation energy. A plot of log r_A vs. $1/T$ for such a reaction at a constant value of P_A is shown in Task 6.2. How would you determine the values of E_a and ΔH_a^o from such a plot?

TASK 6.2 True and Apparent Activation Energies for Eq. (6.4)

The figure shows an Arrhenius plot for a unimolecular catalysed reaction. Explain the shape of the plot.

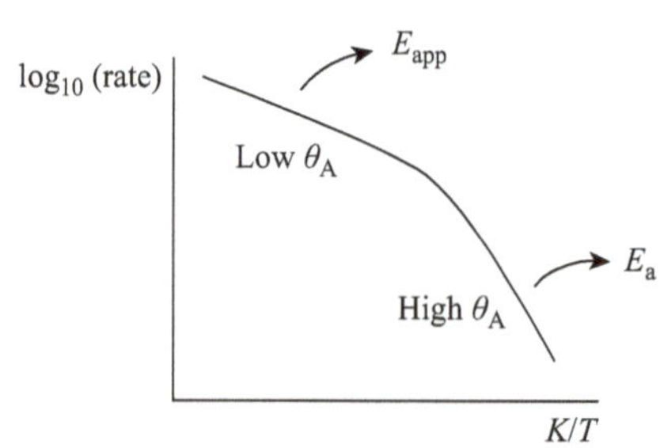

As drawn, the axes of this plot do not show any values. If the reaction rate in the low temperature region increases by a factor of two when the temperature is raised from 50 °C to 70 °C, calculate the true activation energy, E_a, and the enthalpy of adsorption of A on the surface, ΔH_a^o.

Hint: You will have to make an estimate of the gradients of the lines shown at low and high temperatures.

6.4 Bimolecular Reactions — Eley—Rideal Kinetics

The Eley—Rideal mechanism describes the situation which occurs in a bimolecular reaction when one of the reactants interacts directly from the gas phase with an adsorbed layer of the other reactant:

$$A_{(g)} + B_{(ads)} \rightarrow P_{(g)} \tag{6.17}$$

The mechanism was first postulated by Eley and Rideal in 1939[8] in relation to work on hydrogen reactions on metals but is more usually nowadays applied to more complicated situations. In essence, the mechanism is based on the idea that one reacting species (e.g. *H*) is adsorbed on the surface at effectively full coverage and that the other ($D_{2(g)}$) reacts with it from the gas phase (perhaps as a physically adsorbed species) to form the product:

$$D_{2(g)} + H_{(ads)} \rightarrow HD_{(g)} + D_{(ads)} \tag{6.18}$$

The kinetic expression is set up by using a Langmuir expression for the concentration of $B_{(ads)}$ and using P_{A} to express the concentration of *A*:

$$\mathrm{Rate} = -\mathrm{d}P_{\mathrm{A}}/\mathrm{d}t = k_{\mathrm{T}} \cdot P_{\mathrm{A}} \cdot \theta_{\mathrm{B}} = k_{\mathrm{T}} \cdot P_{\mathrm{A}} \cdot b_{\mathrm{B}}P_{\mathrm{B}}/(1 + b_{\mathrm{B}}P_{\mathrm{B}}) \tag{6.19}$$

It is interesting to note that an equivalent equation could be set up if it was assumed that *A* and *B* adsorbed on different sorts of sites so that two separate Langmuir isotherms can be used to express the rate:

$$\mathrm{Rate} = -\mathrm{d}P_{\mathrm{A}}/\mathrm{d}t = k_{\mathrm{T}} \cdot \theta_{\mathrm{A}}\theta_{\mathrm{B}} = k_{\mathrm{T}} \cdot \{b_{\mathrm{A}}P_{\mathrm{A}}/(1 + b_{\mathrm{A}}P_{\mathrm{A}})\} \cdot \{b_{\mathrm{B}}P_{\mathrm{B}}/(1 + b_{\mathrm{B}}P_{\mathrm{B}})\} \tag{6.20}$$

This reduces to Eq. (6.17) if $b_{\mathrm{A}}P_{\mathrm{A}} << 1$; in other words, if *A* is weakly adsorbed.

6.5 The Mars—Van Krevelen Mechanism

The third commonly encountered mechanistic approach of importance in catalysis is the Mars—Van Krevelen mechanism. This mechanism is generally used to describe the kinetics of reactions such as selective oxidation of hydrocarbons-although it was first developed to explain the kinetics of the oxidation of SO_2 over a molten *V/K* salt supported on a porous silica support.[9] It has more recently been applied to many other types of reaction such as hydrodesulfurization and NO_x removal.[10]

---

8 D.D. Eley was at the time a research student working with Sir Eric Rideal in Cambridge on ortho—para hydrogen exchange over metal surfaces. The mechanism has also been named the Langmuir—Rideal mechanism as it has been suggested that Langmuir postulated the mechanism as early as 1921 (http://www.answers.com/topic/langmuir-rideal-mechanism\).

9 P. Mars, D.W. van Krevelen. Chem. Eng. Sci. Spec. Suppl. 3 (1954), p. 41.

10 An excellent review of more recent applications of the mechanism is given in an article by C. Doornkamp, V. Ponec, J. Molec. Catal. A: Chem. 162 (2000) 19—32.

The Mars-van Krevelen mechanism is based on the idea that adsorption of one molecule occurs on top of another molecule which had previously been adsorbed.[11] We will now derive the appropriate equation for a very general case, the all-over reaction:

$$A_{(g)} + B_{(g)} \rightarrow C_{(g)} \tag{6.21}$$

We assume that the reaction takes place according to the following steps, where □ represents an empty site on the surface of the catalyst:

$A_{(g)} + B_{ads} \rightarrow A - B_{ads}$	(a) Adsorption of A on top of adsorbed B	(6.22)
$A - B_{ads} \rightarrow C_{ads}$	(b) Reaction of adsorbed complex	
$C_{ads} \rightarrow C_{(g)} + \square$	(c) Desorption of product C	
$B_{(g)} + \square \rightarrow B_{ads}$	(d) Adsorption of B	

The adsorption of B (step d) is shown last since the empty site is created by the reaction rather than being present on the fresh catalyst.[12] In setting up a kinetic relationship, steps (a) and (b) are combined, assuming that the product is formed as soon A interacts with an adsorbed B. It is also assumed that the desorption step (c) is rapid compared with either this combined reaction or step (b). We then assume that the reaction effectively occurs in only two steps: the adsorption of B with rate constant k_2 (step (d)); and a second (a combination of all the other steps), the formation of an empty site as well as the products by the reaction:

(e) $A_{(g)} + B_{ads} \xrightarrow{k_1} C_{(g)} + \square$

If the coverage of the surface by adsorbed B species is given by θ_B and that of empty sites is given by $\theta_\square$ (where $\theta_B + \theta_\square = 1$), the rates of these two steps are given by:

$$r_1 = k_1 \cdot P_A \cdot \theta_B \quad \text{and} \quad r_2 = k_2 \cdot P_B \cdot \theta_\square = k_2 \cdot P_B \cdot (1 - \theta_B) \tag{6.23}$$

from which, with $r_1 = r_2$ in a steady state, we obtain

$$\theta_B = k_2 P_B / (k_1 P_A + k_2 P_B) \tag{6.24}$$

Hence the rate of reaction is given by:

$$\text{Rate,}\ r = k_1 \cdot P_A \cdot \theta_B = k_1 k_2 P_A P_B / (k_1 P_A + k_2 P_B) \qquad (6.25)^{13}$$

11 The original derivation was for the adsorption of SO_2 on top of a pre-adsorbed oxygen atom.
12 In practice, $B_{(g)}$ is often oxygen and so we then also have to consider the dissociation of the O_2 molecule.
13 It is interesting to note that the *form* of the Mars–Van Krevelen Eq. (6.25) is very similar to that of the Eley–Rideal equation for the reaction of A(g) from the gas phase with a surface containing both A_{ads} and B_{ads} if one assumes that the adsorption of both molecules is strong (the 1 in the denominator can be ignored) but that A does not react with B on the surface to any appreciable extent.

BOX 6.3 Application of the Mars–Van Krevelen Mechanism to the Selective Oxidation of a Hydrocarbon

Let us assume that the reaction occurs by two steps:

$$C_xH_y + O_{ads} \xrightarrow{k_1} \text{Oxidation products} + \square \qquad \text{(Reduction)} \qquad (1)$$

and

$$0.5O_2 + \square \xrightarrow{k_2} O_{ads} \qquad \text{(Oxidation)} \qquad (2)$$

The concentration of the active oxygen on the catalyst surface is determined, as in the derivation in the main text, by the relative rates of reactions 1 and 2. We assume here for simplicity* that one oxygen atom is needed to produce the (selective) oxidation products from the hydrocarbon (HC) and that there is no parallel non-selective oxidation process to H_2O and CO_2.

Following the same arguments as for the general case in the main text (but now with the rate of step (2) expressed as $k_2P_{O_2}^{1/2}(1 - \theta_O)$, where θ_O is the coverage of the surface by oxygen species), we obtain:

$$\text{Rate} = r = k_1k_2[\text{HC}]P_{O_2}^{1/2}/\{k_1[\text{HC}] + k_2P_{O_2}^{1/2}\} = 1/\{1/k_1[\text{HC}] + 1/k_2P_{O_2}^{1/2}\} \qquad (3)$$

or

$$1/r = 1/k_1[\text{HC}] + 1/k_2P_{O_2}^{1/2} \qquad (4)$$

There are two limiting cases as before:

$$r_{red} = k_1 \cdot P_{HC} \qquad (5)$$

and

$$r_{ox} = k_2 \cdot P_{O_2}^{1/2} \qquad (6)$$

Taking as an example the oxidation of a hydrocarbon over a V_2O_5-containing catalyst for which the activity is due to the oxidation/reduction cycle:

$$V^{5+} + e \rightleftarrows V^{4+} \qquad (7)$$

Eq. (4) would apply to the situation in which the active sites were full of oxygen (V^{5+}) and Eq. (5) would apply if the sites were fully reduced (V^{4+}).

* The conventional derivation has recently been criticized by Vannice (Catal. Today 123 (2007) 18–22) who argues that, although the redox model may apply, the mathematical treatment is acceptable only for the adsorption of molecular oxygen on a single site. He argues that the redox model is better described by a Langmuir–Hinshelwood or Hougen–Watson approach. The approach described above works as long as step (2) is not an elementary step but encompasses adsorption of oxygen followed by dissociation of the oxygen to allow it to be adsorbed on two distinct sites. We discuss kinetic expressions allowing for dissociation of the adsorbed species in the following sections.

which can be rewritten in a much simpler form as:

$$1/r = 1/k_1 P_A + 1/k_2 P_B \tag{6.26}$$[14]

There are two limiting cases:

If $k_1 << k_2$, then

$$r = k_1 P_A \tag{6.27}$$

and if $k_2 << k_1$, then

$$r = k_2 P_B \tag{6.28}$$

The Mars–Van Krevelen equation is most frequently encountered in relation to selective oxidation reactions. We treat the example of the selective oxidation of a hydrocarbon in Box 6.4.

6.6 Practical Examples of Mechanistic Kinetic Expressions

6.6.1 The Langmuir Equation and the Pressure Function

We will now focus our attention on some examples of kinetic expressions that have been set up for reactions of practical importance. In the examples of the three main sorts of mechanism given above, we have considered only abstract molecules, A, B, HC, etc. We will now consider how the equations are modified if dissociative adsorption of one or more of the molecules taking part in the reaction occurs. Our first example concerns the kinetics of the synthesis of ammonia.

But first, a brief recap. In Chapter 2, we discussed the formal equations that can be set up using the Langmuir approach for the dissociative adsorption of molecules such as hydrogen and methane (as CH_3 groups) and obtained for these two molecules the following isotherms:

$$\theta_H = C_H P_{H_2}^{0.5} \Big/ \left(1 + C_H P_{H_2}^{0.5}\right) \tag{2.11}$$

which for intermediate coverages simplifies to:

$$\theta_H = C_{H^n} P_{H_{2n}} \tag{2.12}$$

and

$$\theta_{CH_3} = k P_{CH_4} / P_{H_2}^{0.5} \Big/ \left(1 + k P_{CH_4} / P_{H_2}^{0.5}\right) \tag{2.14}$$

[14] This is equivalent to the equation giving the total resistance for two electrical resistances in series. It can be extended further by adding an equivalent reciprocal term for the rate of diffusion of gaseous molecules to the surface (external diffusion). We will discuss external diffusion limitations in Chapter 7. It is interesting to note that an equivalent term for internal diffusion *cannot* simply be added as the expressions allowing for internal diffusion include terms arising from the chemical kinetic expressions.

BOX 6.4 Pressure Functions

If we wish to set up a Langmuir adsorption expression for a species such as $C_2H_{5(ads)}$ resulting from the adsorption of ethane on a uniform surface, we may consider the following equilibrium occurs:

$$C_2H_{6(g)} \rightleftarrows C_2H_{5(g)} + 1/2H_{2(g)} \tag{1}$$

and then write:

$$K_{C_2H_5} = P_{C_2H_5} P_{H_2^{1/2}} / P_{C_2H_6} \tag{2}$$

Hence, we can write:

$$P_{C_2H_5} = K_{C_2H_5} P_{C_2H_6} / P_{H_2}^{1/2} \tag{3}$$

This is known as a "pressure function" and expression (3) can be inserted in a Langmuir adsorption isotherm to give the appropriate isotherm for the adsorption of C_2H_5 species*:

$$\begin{aligned}\theta_{C_2H_5} &= b_{C_2H_5} \cdot P_{C_2H_5} / (1 + b_{C_2H_5} P_{C_2H_5}) \\ &= b_{C_2H_5}\, K_{C_2H_5} \cdot P_{C_2H_6} / P_{H_2}^{1/2} / (1 + b_{C_2H_5} K_{C_2H_5} \cdot P_{C_2H_6} / P_{H_2}^{1/2}) \\ &= c\, P_{C_2H_6} / P_{H_2}^{1/2} / (1 + c\, P_{C_2H_6} / P_{H_2}^{1/2})\end{aligned} \tag{4}$$

where the constant *c* incorporates both the equilibrium constant and the adsorption constant.

For intermediate coverages, Eq. (4) can be simplified further to give a power expression:

$$\theta_{C_2H_5} = c'\, P^n_{C_2H_6} / P^{n/2}_{H_2}$$

This treatment assumes that only C_2H_5 species are adsorbed on the surface to any appreciable extent and that all the reactions up and to the adsorption of the C_2H_5 species are in equilibrium. Similar expressions can be set up for any such species as long as similar assumptions can be made. We will see below that such an approach has been made in the case of the kinetics of ammonia synthesis.

* See C. Kemball, Thermodynamic factors in adsorption and catalysis: equilibria in the adsorbed phase, Disc. Faraday Soc. 41 (1966) 190.

which again simplifies to

$$\theta_{CH_3} = k' (P_{CH_4} / P^{0.5}_{H_2})^n \tag{2.15}$$

where $0 < n < 1$.

Similar power expressions can be written (see also Box 6.4) for many other adsorbed species resulting from dissociative adsorption. If we are concerned with a reaction involving dissociated species such as $CH_{3(ads)}$ or $H_{(ads)}$, then it is possible with such expressions to set up kinetic expressions which can be considered to have some basis in mechanism, even though the resultant expressions could also be considered to be semi-empirical rate expressions. We will now deal with several examples of detailed kinetic studies reported in the literature in which such relationships are used to construct mechanistic kinetic rate laws. The reader is encouraged to follow up on some of the examples given below and to search the literature for other examples.

6.6.2 Ammonia Synthesis

We have discussed the importance of the catalytic synthesis of ammonia by the Haber—Bosch Process in some detail in Chapter 1 in which we saw that the recognition that this reaction could be carried out, as well as the development of the catalysts used and the process itself, were all important landmarks in the modern application of catalysis. Ammonia synthesis occurs by the reaction:

$$N_{2(g)} + 3H_{2(g)} \rightleftarrows 2NH_{3(g)} \tag{6.29}$$

As we will see in Chapter 7, the reaction is severely limited by thermodynamic considerations and so the reverse reaction occurs to a significant extent as soon as a reasonable conversion is reached. Many papers have been written on the subject of the kinetics of ammonia synthesis and many consider the effects of the reverse reaction as well as the forward reaction. Here we will consider only the forward reaction in any detail; the interested reader should consult one of the more advanced textbooks available[15] to find more details of fuller treatments, including a discussion of the idea of stoichiometric number as applied to multistep reactions.

We now consider the derivation of an expression for the forward reaction to explain the kinetics of the reaction over an iron catalyst at very low conversions. This expression was originally proposed by Temkin and Pyzhev (Acta Physicochim. USSR, 12 (1940) 3270) on the basis of the Elovich equation (which allows for a variation of the energetics of adsorption with coverage) but the approach was modified by Ozaki et al. (Proc. Royal Soc. A258 (1960) 47), by using a Langmuir approach with the assumption of a uniform surface. In common with the earlier approach of Temkin and Pyzhev, these authors assumed that the rate determining step in the synthesis reaction is the rate of the dissociative adsorption of nitrogen on a nitrogen-covered surface:

$$r = k_a P_{N_2}(1 - \theta_N)^2 \tag{6.30}$$

The value of θ_N is given by a Langmuir isotherm of the form:

$$\theta_N = bP_N^*/(1 + bP_N^*) \tag{6.31}$$

where P_N^* is the effective partial pressure of nitrogen atoms (sometimes known as the "virtual pressure" of the nitrogen atoms[16]) that would be in equilibrium with the adsorbed nitrogen atoms. As this virtual pressure cannot be measured directly, it is calculated assuming that they are present in equilibrium according to:

$$N^*_{(g)} + 1.5H_{2(g)} \rightleftarrows NH_{3(g)} \tag{6.32}$$

---

[15] See, for example, J.M. Thomas, W.J. Thomas, Introduction to the Principles of Heterogeneous Catalysis, Academic Press, London and New York (1967), or C.H. Bartholomew, R.J. Farrauto, Fundamentals of Industrial Catalytic Processes, 2nd Edition, Wiley Interscience, New Jersey, 2006.

[16] The virtual pressure is equivalent to the pressure function referred to in Box 6.4.

for which the equilibrium constant is given by:

$$K_N^* = P_{NH_3}/P_N^* \cdot P_{H_2}^{1.5} \tag{6.33}$$

Inserting Eq. (6.33) in Eq. (6.31) gives:

$$\theta_N = K \cdot P_{NH_3}/P_{H_2}^{1.5}/\{1 + K \cdot P_{NH_3}/P_{H_2}^{1.5}\} \tag{6.34}$$

where $K = b/K_N^*$

Substituting Eq. (6.32) in Eq. (6.30), after slight manipulation (check!), gives:

$$r = k' P_{N_2}/\{1 + K \cdot P_{NH_3}/P_{H_2}^{1.5}\}^2 \tag{6.35}$$

This equation, which can be approximated by:

$$r = k_a P_{N_2} \ \{P_{H_2}^{1.5}/P_{NH_3}^2\}^n$$

where $0 < n < 1$, gives a reasonable description of the kinetics at low conversions of nitrogen to ammonia. At higher conversions, the rate of the reverse reaction becomes appreciable and so, as noted above, a term has to be added for the reverse reaction. The full Temkin–Pyzhev equation that accounts for the rates of the forward and reverse reactions is:

$$r = k' P_{N_2}\{P_{H_2}^3/P_{NH_3}^2\}^m - k''\{P_{NH_3}^2/P_{H_2}^3\}^{(1-m)} \tag{6.36}$$

It should be recognized that when $r = 0$, the reaction is at equilibrium and then

$$k' P_{N_2}\{P_{H_2}^3/P_{NH_3}^2\}^m = k''\{P_{NH_3}^2/P_{H_2}^3\}^{(1-m)} \tag{6.37}$$

or

$$k''/k'' = P_{NH_3}^2/(P_{N_2} \cdot P_{H_2}^3) = K_{eq} \tag{6.38}$$

BOX 6.5 Ammonia Decomposition

It is much easier to study ammonia decomposition than ammonia synthesis since the decomposition reaction, when carried out at reasonably high temperatures, can be considered to be essentially irreversible. Further, it need not be studied under high pressure and so the interpretation of kinetic measurements is much more straightforward. Nevertheless, unless there is a significant change in the reaction mechanism with change in temperature, the kinetic parameters determined for the decomposition reaction should be directly related to those for the forward (synthesis) reaction. This feature is particularly useful in making comparisons between various different catalyst compositions. It is interesting to note that there has recently been a renewed interest in the catalytic decomposition of ammonia as a source of clean hydrogen for fuel cell applications.

TASK 6.3

Use the approach outlined above to derive a general power rate kinetic expression for the decomposition of ammonia over a metal catalyst assuming that combination of two nitrogen atoms is the rate-determining step.

TASK 6.4

Examine the recent literature on the mechanism if the synthesis and decomposition of ammonia. Start by searching in Scopus for the following paper: J.A. Dumesic, A.A. Trevino, J. Catal. 116 (1989) 119. From this, search for more up to date references, particularly review articles.

6.6.3 Ethane Hydrogenolysis

Much work has been reported in the literature on the mechanism of the conversion of hydrocarbons over metal catalysts, both unsupported (e.g. metallic films) and supported. This work is related to a variety of different reactions: hydrogenation, hydrogenolysis, isomerization and cracking/hydrocracking (see Chapter 8). Common to all these reactions is the adsorption of the hydrocarbon on a clean metallic surface, this involving as a first step dissociation of a C−H bond and adsorption of the hydrocarbon species, together with hydrogen atoms, on the metal surface. We will now discuss the model set up by J.H. Sinfelt for the hydrogenolysis of ethane to give methane. His analysis is based on the following steps occurring on the metal surface:

$$\begin{aligned} &C_2H_{6(g)} \rightleftarrows C_2H_{5(ads)} + H_{(ads)} && \text{(i)} \\ &C_2H_{5(ads)} + H_{(ads)} \rightleftarrows C_2H_{x(ads)} + \alpha H_{2(g)} && \text{(ii)} \\ &C_2H_{x(ads)} \rightarrow \text{adsorbed } C_1 \text{ fragments } (+H_{(ads)} \rightarrow CH_{4(g)}) && \text{(iii)} \end{aligned} \tag{6.39}$$

The ethane molecule adsorbs by the initial dissociation of a C−H bond and the $C_2H_{5(ads)}$ species formed then dehydrogenates rapidly and reversibly to give the species $C_2H_{x(ads)}$; this is therefore in equilibrium with the gaseous ethane. The quantity α is equal to $(6 - x)/2$. Sinfelt's analysis assumes that the $C_2H_{x(ads)}$ species is the only one present on the surface in any appreciable concentration and that it reacts in a slow step to give adsorbed C_1 fragments (step iii), these then giving $CH_{4(g)}$ by rapid interaction with adsorbed hydrogen species.

The coverage of the surface by $C_2H_{x(ads)}$, θ_x, is then given by Eq. (6.40):

$$\theta_x = (KP_E/P_{H_2}^{\alpha})/(1 + KP_E/P_{H_2}^{\alpha}) \tag{6.40}$$

Simplifying this expression for intermediate coverages:

$$\theta_x = (KP_E/P_{H_2}^{\alpha})^n \tag{6.41}$$

where, as before, $0 < n < 1$.

The rate of hydrogenolysis to methane is then given by that of step (6.39.iii):

$$r = k_{iii} \cdot \theta_X = k_{iii} \cdot K^n \cdot (P_E / P_{H_2}^{\alpha})^n = k P_E^n P_{H_2}^{(-n\alpha)} \tag{6.42}$$

Sinfelt had previously reported[17] kinetic measurements for the hydrogenolysis of ethane over a range of different metals supported on a silica support and he fitted those results to Eq. (6.42) as shown in Table 6.1.

Table 6.1: Kinetic of ethane hydrogenolysis. Analysis of hydrogen pressure dependencies for different metals (J. Catal. 3 (1969) 175).

				Exponents	In Rate Experimental	Expression
Catalyst	Temperature/°C	x	α	Ethane (n)	Hydrogen (m)	Hydrogen ($-n\alpha$)
Co	219	4	1	1.0	−0.8	−1.0
Ni	177	2	2	1.0	−2.4	−2.0
Ru	188	2	2	0.8	−1.3	−1.6
Rh	214	0	3	0.8	−2.2	−2.4
Pd	354	0	3	0.9	−2.5	−2.7
Os	152	2	2	0.6	−1.2	−1.2
Ir	210	2	2	0.7	−1.6	−1.4
Pt	357	0	3	0.9	−2.5	−2.7

Reproduced with kind permission of Elsevier.

As noted above, the quantity x denotes the number of hydrogen atoms associated with the C_2 species on the surface before dissociation occurs. It can be seen that Co gives the highest value of x while Pd and Pt give rise to complete removal of hydrogen in the adsorbed species. Table 6.1 also shows the experimentally determined exponents for ethane (n) and hydrogen (m) and the calculated values for hydrogen ($-n\alpha$) for each of the metals. The agreement between the experimental and theoretical exponents is very close, indicating that the model works very well for the conditions studied.[18]

TASK 6.5

Examine using Scopus or Web of Science the articles that have cited the paper by Sinfelt described above (see footnote 17). Pay particular attention to the paper by Rostrup Nielsen and Alstrop on the relationship between hydrogenolysis and steam reforming. Follow up on the work from the Topsøe laboratories on the effects of sulfur and gold on the steam reforming reaction.

---

[17] J.H. Sinfelt, Catal. Rev. 3 (1969) 175.

[18] In his paper cited above, Sinfelt goes on to discuss the effect of carrying out the hydrogenolysis reaction at higher temperature and shows that the dependence of the rate on hydrogen partial pressure is temperature dependent at these temperatures.

6.6.4 Methanation

The subject of the methanation of CO was the subject of much attention in the period around 1980 and it was one of the top 15 "research fronts" in the first 6 months of 1982.[19] This was largely because there was a shortage of natural gas, particularly in the northern states of the USA, and there was a renewed interest in the possibility of producing natural gas from syngas made by coal gasification.[20] Many papers on the subject were published during that period, these discussing topics varying from the fundamentals of the mechanism of the reaction to plant operation. We will return to one aspect of the operation of full-scale reactors for this reaction in Chapter 7. However, here we will concentrate on the kinetic aspects of the reaction.

In 1975, Vannice published two papers[21] in which he described the measurement of the kinetics of the methanation reaction over a series of Group A metals and fitted his rate data to a kinetic expression:

$$\text{Rate} = k \cdot P_{H_2}^{x} \cdot P_{CO}^{y} \tag{6.43}$$

He derived an equivalent theoretical equation on the assumption that the rate determining step in the reaction was the interaction of adsorbed hydrogen atoms with CHOH species adsorbed on the surface:

$$CHOH_{(ads)} + y/2H_{2(ads)} \rightarrow CH_{y(ads)} + H_2O_{(g)} \tag{6.44}$$

Ollis and Vannice later published[22] a more satisfactory treatment based on the rate determining step shown in Eq. (6.44) in which they derived a Langmuir expression for $CHOH_{(ads)}$:

$$\theta_{CHOH} = K \cdot P_{CO} \cdot P_{H_2} / (1 + KP_{CO}P_{H_2}) \tag{6.45}$$

which was approximated by:

$$\theta_{CHOH} \approx (K \cdot P_{CO} \cdot P_{H_2})^n \tag{6.46}$$

with $0 < n < 1$ as in the previous derivations. Their resulting theoretical equation had the form:

$$\text{Rate} = k\{P_{CO} \cdot P_{H_2} \cdot (P_{H_2}/P_{CO})^{y/2}\}^{1/2} \tag{6.47}$$

19 E. Garfield, Current Contents (Physical and Chemical Sciences), 22 (1982) 5.

20 For a review of methanation and steam reforming, the reader is referred to the following: J.R.H. Ross, Metal catalysed methanation and steam reforming, Catalysis 7 (1985), 1–45, Special Periodical Reports, Royal Society of Chemistry. (Available for download from www.rsc.org/ebooks/archive/free/.../CL9780851865843-00001.pdf\).

21 M.A. Vannice, The catalytic synthesis of hydrocarbons from H_2/CO mixtures over the group VIII metals, Parts 1 and 2, J. Catal. 37 (1975) 449–461 and 462–473.

22 D.F. Ollis, M.A. Vannice, J. Catal. 38 (1975) 514.

in which k incorporates several constants and y is the number of hydrogen atoms involved in the rate determining step (Eq. (6.44)).

Later work on the methanation reaction showed quite clearly that the reaction proceeds not through $CHOH_{(ads)}$ species but through adsorbed carbon atoms derived from the dissociative adsorption of CO. The present author derived an expression for the kinetics of the methanation reaction making the assumption that the following equilibria occur[23]:

$$\begin{aligned} &CO_{(g)} \rightleftarrows CO_{(ads)} && \text{(i)} \\ &CO_{(ads)} \rightleftarrows C_{(ads)} + O_{(ads)} && \text{(ii)} \\ &H_{2(g)} \rightleftarrows 2H_{(ads)} && \text{(iii)} \\ &2H_{(ads)} + O_{(ads)} \rightleftarrows H_2O_{(g)} && \text{(iv)} \end{aligned} \tag{6.48}$$

with the following reaction as rate determining:

$$C_{(ads)} + yH_{(ads)} \rightarrow CH_{y(ads)} \tag{v}$$

The species $CH_{y(ads)}$ then combines rapidly with more hydrogen species and methane is desorbed.

The resultant rate expression is:

$$\text{Rate} = k\cdot\frac{P_{CO}P_{H_2}^{(1+y/2)}/P_{H_2O}}{(1+b_cP_{co}P_{H_2}/P_{H_2O})^{1+y}} \tag{6.49}$$

Multiplying by $P_{CO}P_{H_2}/P_{H_2O}$, one obtains:

$$\text{Rate}\ \alpha\ \frac{\{P_{co}P_{H_2}/P_{H_2O}\}^{(1+y)}P_{H_2O}^{y}}{(1+b_cP_{co}P_{H_2}/P_{H_2O})^{1+y}P_{CO}^{y}P_{H_2}^{y/2}} \tag{6.50}$$

Making the same approximation as above, the resulting equation is:

$$\text{Rate}\ \alpha\ P_{CO}^{\{n+y(n-1)\}}\ P_{H_2O}^{-\{n+y(1-n)\}}$$

The most significant difference between this expression and that obtained by Ollis and Vannice is the inclusion for a term for the water partial pressure. Unfortunately, no work in which the effect of water vapour on the kinetics of the methanation reaction has been published; some work carried out at the time using a microbalance system with gas analysis of the exit gases gave strong support for this kinetic expression[24] but it was never possible to obtain sufficiently good control of the water partial pressures to obtain unequivocal results.

[23] J.R.H. Ross, A modified kinetic expression for the methanation of carbon monoxide over Group VIII metal catalysts, J. Catal. 71 (1981) 205.

[24] B. Höhlein, R. Menzer, J.R.H. Ross, J. Sarkar, unpublished results.

TASK 6.6

R.Z.C. van Meerten et al. (Appl. Catal. 3(1982) 29) have carried out a painstaking examination of the kinetics of methanation over a nickel/silica catalyst. Study the results of that paper and discuss the model put forward.

Then carry out a Scopus or Web of Science search on this paper and follow up some of the most recent papers citing this work, paying particular attention to several other papers listed which have J. Rostrup Nielsen as a co-author.

CHAPTER 7

Large-Scale Catalytic Reactors

Chapter Outline

7.1 Introduction

In Chapter 5, we considered the types of reactors used in laboratory studies to investigate catalytic reactions. In such studies, the work is usually done with small catalyst particles and the reaction conditions are chosen to avoid diffusion problems. As the reactors become larger, larger catalyst geometries are used (see below) and it becomes impossible to ignore problems introduced as a result of mass and heat transfer. A detailed description of the design and operation of industrial catalytic reactors is beyond the scope of this book and so we will restrict ourselves largely to the limitations placed on kinetic measurements in larger reactors by mass transfer restrictions and to a brief discussion of some of the more important types of reactors used, particularly in relation to reactions operating close to thermodynamic equilibrium. We will start with a brief discussion of the forms of catalyst needed for full-scale reactors and will then discuss in turn the problems associated with external and internal diffusion. We will then examine a number of examples of the reactors used for catalytic processes, particularly ones in which heat and mass transfer are of importance. Subsequent chapters will then show how some of the topics which have been covered in the preceding chapters have been applied in practice.

Heterogeneous Catalysis. DOI: 10.1016/B978-0-444-53363-0.10007-6

Figure 7.1:
Some typical catalyst geometries used in full-scale reactors.

7.2 Catalyst Geometries

Some typical examples of particle geometries used in commercial reactors are shown in Figure 7.1, these being made from the appropriate powders by techniques such as granulation, pelleting and extrusion. The objective is to create particles with large accessible external surface areas that have adequate strength so that they do not fragment in the reactor, particularly during the charging process, such fragmentation being responsible for high back pressures. The catalyst must also have long operating lives (often of the order of years of continuous operation) and a good resistance to poisoning and sintering. However, first and foremost, it is essential in almost all cases that the catalyst particles have adequate pore geometries to enable the reactants to diffuse to the catalyst surface and for the products to leave the surface.

7.2.1 Pelleting

We will here consider only the cases of catalyst pelleting and extrusion. The pelleting process is similar to that used in the forming of pharmaceutical tablets: the appropriate powder[1] is added to a cylindrical cavity and pressure is applied above and below by pistons. The tablet is then released from the cavity and the procedure is repeated. The process is fully automated and typical tableting presses contain a large number of such cavities and pistons arranged around the circumference of a circular plate as depicted schematically in Figure 7.2. Often, to enable the powder to flow properly, lubricants are added to the powder either before the process is started or as the powder is added to the die, these additives including surface active agents such as stearates or even graphite. In some cases, other materials are also added to give the pellets formed extra strength. The material is then frequently "calcined" to remove any lubricant materials added during the pressing step. As a general rule, the strength of the final pellet increases as the applied pressure during pelleting is increased: necks form between the catalyst particles and these grow larger with increased time of pressing and applied pressure. The calcination process may also add to the strength of the pellet, especially if the precursor powder had not been calcined.

[1] The pelleting process can be applied either to the catalyst carrier alone or to the final formulation; in the case that only the carrier is pelleted, the active component is added to the preformed carrier by impregnation as discussed in Chapter 4, Section 4.3.

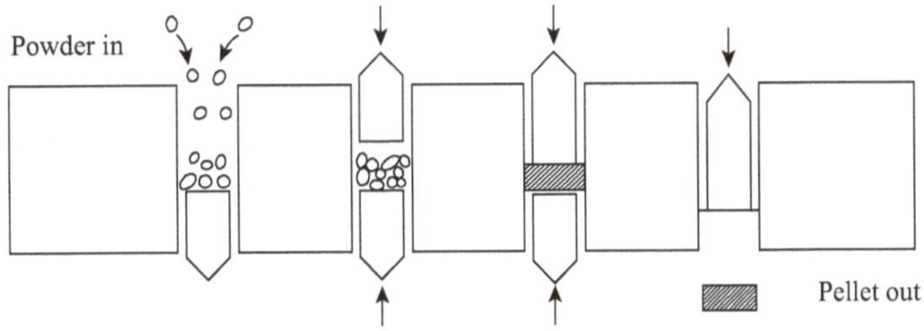

Figure 7.2:
Schematic representation of the pelleting process.

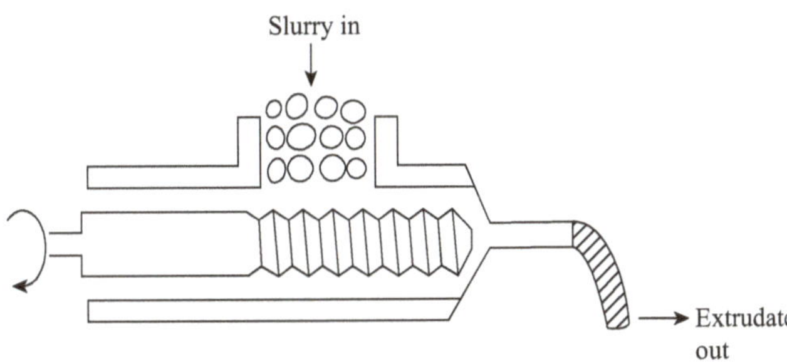

Figure 7.3:
Schematic representation of the extrusion process.

7.2.2 Extrusion

The extrusion process is similar to that encountered when squeezing toothpaste from a tube or mincing meat with a mincer (see schematic representation in Figure 7.3): a slurry of the desired material is added to a hopper at the top and is then pressed through a nozzle of the desired shape using a screw-shaped central cylinder. Again, the extrusion properties of the slurry can be improved by the addition of surfactants, etc. The shape of the extrudate can be varied by using different dies; for example, Raschig rings and tri-lobes can be formed as well as ribbons and solid cylinders. Once formed, the extrudates must be dried carefully and they are then generally calcined to improve their strength.

BOX 7.1 Ni/Al/La/α-Al_2O_3 Catalysts

In Chapter 4 (Section 4.4), we discussed the preparation of hydrotalcite precursors in the pores of α-Al_2O_3, the high-temperature form of alumina (see Fig. 4.3). α-Al_2O_3 is generally a dense ceramic material with a very low surface area ($<< 1\ m^2g^{-1}$) and is seldom used as a catalyst support. For the purposes of making the Ni/Al/La materials, the α-Al_2O_3 matrix is formed by an extrusion process in which large pores are built into the structure by incorporating organic fibres of the desired dimensions into the paste used for extrusion through a dye giving hollow-spaghetti-type tubes. These are then sliced into the desired lengths and calcined at high temperatures to burn out the organic fibres and to form the α-Al_2O_3 matrix used to incorporate the

(Continued)

BOX 7.1 Ni/Al/La/α-Al_2O_3 Catalysts—*Cont'd*

hydrotalcite phase ($Ni_6(Al+La)_2(OH)_{16}CO_3 \cdot 4H_2O$); this then gives rise, after calcination and reduction, to the active promoted Ni catalyst.

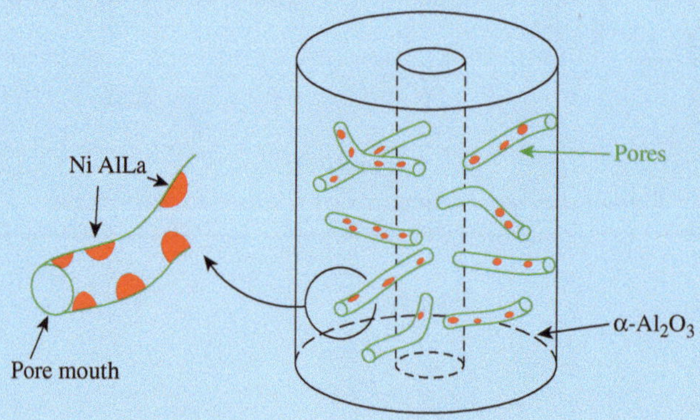

7.3 The Importance of Mass Transfer in Catalysis[2]

Having made a suitable catalyst formulation and having formed the catalyst into a robust shape, we need to know how mass transfer limitations might affect the performance of the catalyst in real reactor geometry. As we discussed in Chapter 4 (see Figure 4.1), a reactant molecule when fed to a reactor must first diffuse towards the external surface of the catalyst pellet and then enter the pore, diffusing through it to the active sites. Once it reacts at these sites, in most cases with another reactant that has also diffused to these sites, the product molecule or molecules must follow the reverse path. Outside the catalyst pellet, the reactants and products diffuse according to the normal laws governing gaseous diffusion and, in the current context, we term this as "external diffusion". Once within the pores, as long as these are relatively narrow, the molecules move according to so-called Knudsen diffusion – in other words, the collisions with the walls are as important as collisions with one another – and we then refer to the occurrence of "internal diffusion". Both diffusion phenomena are of importance in practical applications of catalysis and we will now consider each in turn, paying particular attention to those aspects of greatest importance to those more interested in the fundamental aspects of the subject.

7.3.1 External Diffusion

An important parameter that must be considered in the design of any catalytic reactor, whether of small or large scale, is the linear flow rate of the reactant and product gases

[2] Some of the material in the following sections is derived from a set of lecture notes used by the author when he worked at the University of Twente (NL). These had evolved over the years and contained contributions from H. Bosch, P.J. Gellings, P. Mars, J.G. van Ommen, J. Sonnemans as well as from the author.

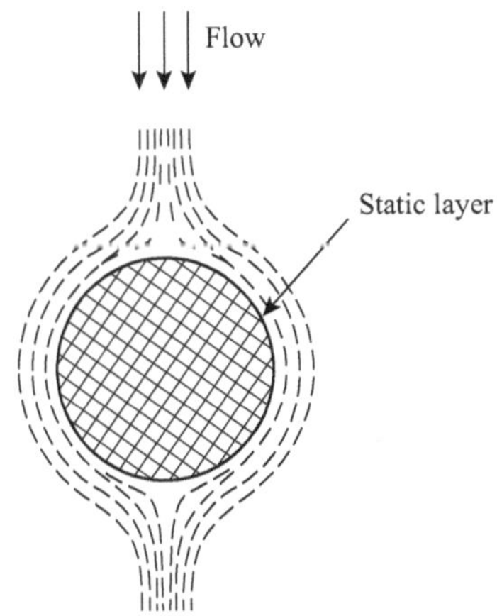

Figure 7.4:
The boundary layer round a spherical particle in a gas flow.

through the reactor. When a gas flows past the boundary of a particle, a static layer of gas molecules builds up alongside the particle and there is gradual transition between this layer and the free-flowing gas phase. This is shown schematically in Figure 7.4 for a spherical particle. (The situation will be somewhat different for irregularly shaped particles but there will still be a static layer surrounded by the free-flowing gas.) The thickness of the boundary layer will depend on a number of factors including the pressure and the composition of the gas flowing over the surface. However, the most important parameter as far as the present discussion is concerned is the gas velocity: the thickness of the boundary layer, and hence the rate of diffusion through it, is inversely proportional to the gas velocity. In other words, if the gas velocity is high enough, the rate of diffusion through the boundary layer at the surface of a catalyst particle at which reaction is occurring will also be high enough to ensure that the concentrations of the reactants and products in the body of the gas will be the same as, or very similar to, that at the surface of the catalyst; under these conditions, "external diffusion" will have no effect on the measured rates. However, if the gas flow rate is too low, the boundary layer thickness will be significant and the rate of diffusion through it will contribute to the observed rates of reaction. The rate of diffusion through such a layer is given by:

$$R_{\mathrm{ext}} = k_{\mathrm{m}}(C_{\mathrm{g}} - C_{\mathrm{s}}) \tag{7.1}$$

where C_g is the concentration of the gas in question in the bulk of the gas, C_s is the concentration of that gas just above the surface of the catalyst and k_m is the mass transfer coefficient. As we will see later, the rate of the surface reaction is then given by some function of C_s. The value of k_m is determined by the physical properties of the gas mixture as well as by the aerodynamic flow conditions and depends on the dimensionless Schmidt (Sc), Reynolds (Re) and Sherwood (Sh) numbers. These numbers are known for most practical situations and so k_m can be worked out. Box 7.2 gives some practical examples of reactions in which external diffusion is important.

BOX 7.2 Examples of External Diffusion Limitations

In the case of catalytic reactions in gas–solid systems, external diffusion generally only plays a significant role in reactions occurring at very high temperatures such as the oxidation of ammonia (10% ammonia in air) over a small number of nets (normally three) composed of Pt/Rh wire of approximately 0.15 mm in diameter. This reaction occurs at 800 °C with a gas velocity through the nets of about 5 ms^{-1}. The conversion obtained is about 98% and the rate is completely controlled by the rate of impact of the reactant molecules with the surface of the wires.

(a)

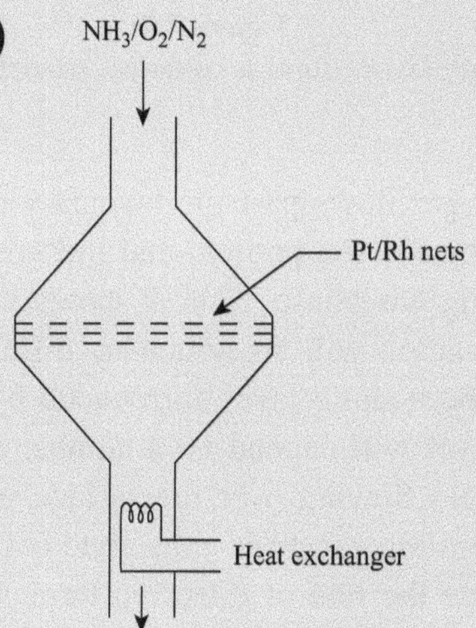

Figure (a) shows the setup used for the oxidation of ammonia. As the reaction is very exothermic, the system includes a heat exchanger downstream of the catalyst gauzes to recover waste heat. In practice, a second set of gauzes at lower temperature is also located below the operating gauzes on which any Pt which evaporates from them is collected by condensation.

A similar situation is encountered in the system used for one of the two processes for the oxidation of methanol to give formaldehyde (http://en.wikipedia.org/wiki/Formaldehyde/) in which the catalyst comprises of a bed of crystallites of silver approximately 1 cm thick sitting on a suitable mesh. The reaction, which is very exothermic, is self-sustaining and operates between 500 and 900 °C with a conversion of methanol of up to 99% and a selectivity to formaldehyde up to 90%. The catalyst life-time can be up to 8 months, depending on the purity of the feed, and it can be regenerated by electrolysis (L. Lefferts, PhD Thesis, U. Twente, 1987).

Another completely different example of mass transfer limitation is that of the hydrogenation of organic compounds (oils or other unsaturated molecules) over nickel catalysts at about 180 °C in a three phase system: gas/liquid/solid. Such a system is shown schematically in Figure (b).

BOX 7.2 Examples of External Diffusion Limitations—*Cont'd*

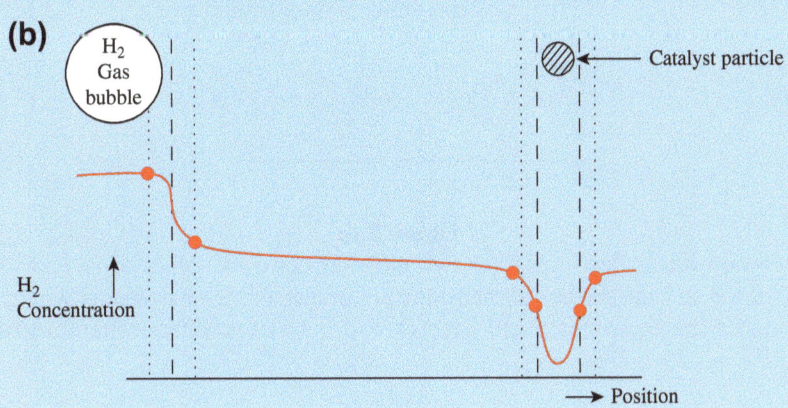

The catalyst particle (suspended in the liquid which is to be hydrogenated) is shown on the right-hand side and a gas bubble is shown on the left. Barriers to the diffusion of hydrogen exist at both interfaces (gas/liquid and liquid/solid) and also throughout the liquid itself. (There is also a concentration gradient within the catalyst particle, as will be discussed further below.) The situation can be much improved by rapid stirring of the system so that the gas bubbles are as small as possible (increasing the gas/liquid interfacial area), thus decreasing the distance for diffusion in the liquid. Stirring also minimizes the gradient at the liquid/solid interface.

In practice, when setting up a set of experiments with a given catalyst system, one should check whether or not external diffusion limitations occur under the chosen conditions. This entails carrying out a series of catalytic conversion measurements in which the linear gas velocity is changed without changing the other conditions of the reaction being studied; in other words, the residence time (or W/F), the temperature and the total bed dimensions must remain constant. Hence, in practice, W must be decreased in sequence by adding an appropriate diluent to keep the bed size the same while varying F at the same time so that W/F remains constant. A typical plot is shown schematically in Figure 7.5; this shows the conversion as a function of flow rate for two different constant values of W/F. (It could also represent results for two different temperatures.) It is clear that the conversion is unaffected by the value of the flow rate (and hence the linear flow velocity) as long as it is above a minimum value and hence external diffusion is unimportant under these conditions. In practice, such measurements are difficult to achieve as there are physical limits to the range of flow rates and catalyst weights that can be used in a particular reaction system.

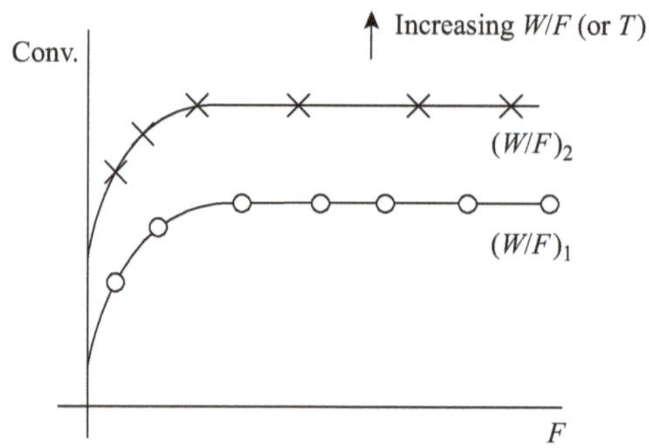

Figure 7.5:
Plots of conversion versus flow rate F for two different constant values of W/F. As W is varied, a diluent is added so that the bed dimensions are unaltered. Similar results would have been obtained if the experiments had been carried out at two different temperatures.

TASK 7.1 External Diffusion

How might you otherwise prove experimentally that a reaction is not affected by external diffusion?

Answer: By varying the reactor geometry while keeping both W and F constant, it is possible to change the linear velocity without changing the value of W/F. For example, by changing the reactor diameter by a factor of two, the linear velocity, being inversely proportional to the reactor cross-sectional area, is changed by a factor of four. If the conversion does not change, then there is no effect of external diffusion.

7.3.2 Internal Diffusion

In Chapter 4 (see Figure 4.1), we indicated the importance of the diffusion of reactants and products into and out of the pores of a catalyst. If this pore diffusion is slow (i.e. we have "internal diffusion limitations"), it can have an effect on the apparent rate of the reaction. If the rate of a catalytic reaction.

$$A + B + \ldots \rightarrow P + Q + \ldots \tag{7.2}$$

without any diffusion limitations is given by:

$$\text{Rate} = k_T \cdot \text{fn}(C_A, C_{B\ldots}, C_P, C_{Q..}), \tag{7.3}$$

where fn(C_A, C_B..., C_P, C_Q..) is a function involving the concentrations of the reactants and products (see Chapter 6), then the reaction rate constant k_T is dependent only on the temperature of the reaction. The dependence of k_T on temperature is given by:

$$k_T = A \cdot \exp\{-E_a/RT\} \tag{7.4}$$

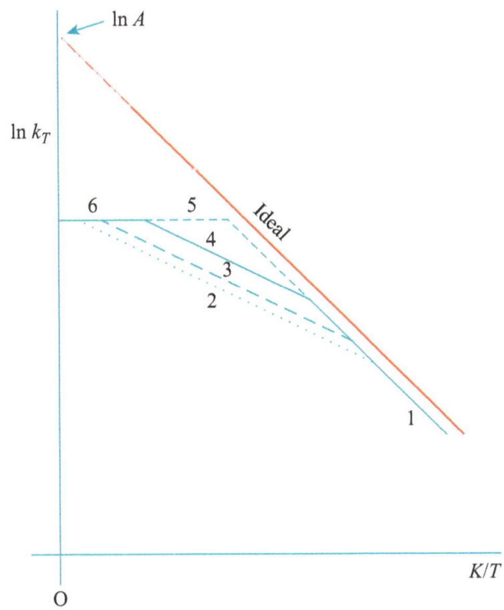

Figure 7.6:
Arrhenius plots showing ideal behaviour (solid line) and deviations due to internal and external diffusion limitations (lines 1–6).

where A is the Arrhenius constant. A plot of $\ln k_T$ vs. $1/T$ will give a straight line of gradient $-E_a/R$ as shown by the solid red line in Figure 7.6. However, if diffusion limitations within pores of the catalyst particle (shown schematically in Figure 4.1) have an effect on the apparent rate of reaction, the plot of $\ln k_T$ vs. $1/T$ deviates from the ideal line (shown by blue lines in Figure 7.6). The deviation from the ideal behaviour is given by the so-called effectiveness factor, η, defined by Eq. (7.5):

$$\eta = \frac{\text{observed reaction rate}}{\text{rate without internal gradients}} = \frac{r_{obs}(C,T)}{r_{chem}(C_s,T_s)} \tag{7.5}$$

where $r_{obs}(C,T)$ is the observed reaction rate (averaged over the whole catalyst particle) at any concentration and temperature and r_{chem} (C_s,T) is the reaction rate at the particle surface at the surface concentration and temperature: in other words, the rate that would occur if no diffusion limitations occurred.

The effectiveness factor is a function of the Thiele modulus, ϕ, a dimensionless number which relates the catalytic activity to the rates of internal mass transfer[3] (see Box 7.3.). The

[3] A deeper treatment of this subject is outside the scope of this book. For further details, the reader should consult a standard text on reactor engineering such as S.H. Fogler, Elements of Chemical Reaction Engineering, Prentice-Hall, London (1986); see also J.A. Moulijn, M. Makkee, A. van Diepen, Chemical Process Technology, Wiley, Chichester, New York, Weinheim, Brisbane, Singapore and Toronto (2001) for an excellent treatment of reactor design.

BOX 7.3 Thiele Modulus, ϕ, and Effectiveness Factor, η

The value of the Thiele modulus, ϕ_{gen}, for an *n*th order reaction is related to k_v, the intrinsic reaction rate constant per unit volume, *n*, the order of reaction, D_{eff}, the effective diffusion rate of the molecules, and C_s, the concentration at the surface of the catalyst particle (pellet, slab, grain, etc.) by the equation:

$$\phi_{gen} = \frac{V}{SA} \cdot \sqrt{\frac{k_v}{D_{eff}} \cdot \frac{n+1}{2} \cdot C_s^{n-1}} \tag{1}$$

The quantity *V*/*SA* is the volume/external surface area ratio of the particle in question. For a first order reaction with a particle of slab geometry (length *L*, with effective diffusivity *Deff*), it can be shown that:

$$\eta = \frac{\tanh\phi}{\phi} \tag{2}$$

where

$$\phi = L\sqrt{\frac{k}{D_{eff}}} \tag{3}$$

relationship between η and ϕ is illustrated for a first order reaction in Figure 7.7. (The figure is given for a plate-shaped catalyst; the shape of the plot is slightly different for cylindrical and spherical particle dimensions.) It can be seen that the effectiveness factor tends to unity at low values of ϕ (high values of D_{eff}; tanhϕ tends to ϕ as ϕ tends to zero) and so all of the catalyst sites

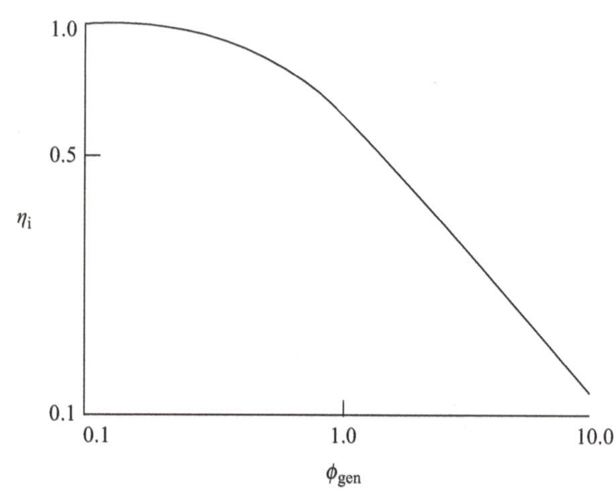

Figure 7.7:
The effectiveness factor, η_i, as a function of the Thiele modulus, ϕ_{gen}, for a first order reaction over a catalyst in the form of a plate.

BOX 7.4 Limiting Values of η

For high values of ϕ, tanh ϕ tends to unity and Eq. 2 of Box 7.3 reduces to:

$$\eta = 1/\phi \tag{1}$$

Hence, the observed reaction rate is now given by:

$$r_{obs} = \eta \cdot r_{chem} = r_{chem}/\phi \tag{2}$$

From Eq. 3 of Box 7.3

$$\phi = \sqrt{k}$$

and so $r_{obs} = r_{chem}/\sqrt{k}$.

For a first order reaction, $r_{chem} = k \cdot (C)$,

$$\therefore r_{obs} = k/\sqrt{k \cdot (C)} = \sqrt{k} \cdot (C) \tag{3}$$

As $k = A \exp(-E/RT)$,

$$r_{obs} = \sqrt{A} \cdot (C) \cdot \exp(-E/2RT) \tag{4}$$

In other words, for a first order reaction, the gradient of the Arrhenius plot is one half of that in a reaction without diffusion limitations and the apparent activation energy $E_{app} = 0.5E_{true}$.

contribute to the measured rates under such conditions. This corresponds to the ideal behaviour (line 1) shown in the Arrhenius plot of Figure 7.6. However, when the temperature increases, η_i decreases and the Arrhenius plot deviates from the ideal line. The temperature at which this deviation occurs depends on many factors, including the catalyst porosity, and the resultant plots are shown as lines 2, 3 and 4. (See Box 7.4 for a further treatment for a first order reaction.) Line 5 shows a situation in which no effect of internal diffusion appears; instead, the line follows the ideal one until the rate reaches a limit corresponding to the rate of mass transfer to the external of the catalyst particles. The same limiting line 6 applies to the other cases (2–4).

Figure 7.8 shows schematically the situation that exists in a spherical catalyst particle containing such pores as a function of reaction temperature. For a low temperature (low ϕ), the concentration profile (line 1) is constant throughout the particle and the effectiveness factor is 1.0. However, as the temperature is increased, the concentration at the centre of the sphere decreases (line 2) and at yet higher temperatures, the concentration at the centre of the sphere can drop to zero (line 3, high ϕ).

An equivalent result is shown more quantitatively in Figure 7.9 which illustrates for an ideal cylindrical pore of length $2L$ open at both ends (e.g. a pore traversing a catalyst pellet) the profile of the concentration of a reactant (plotted as C/Co) versus the distance into the pore

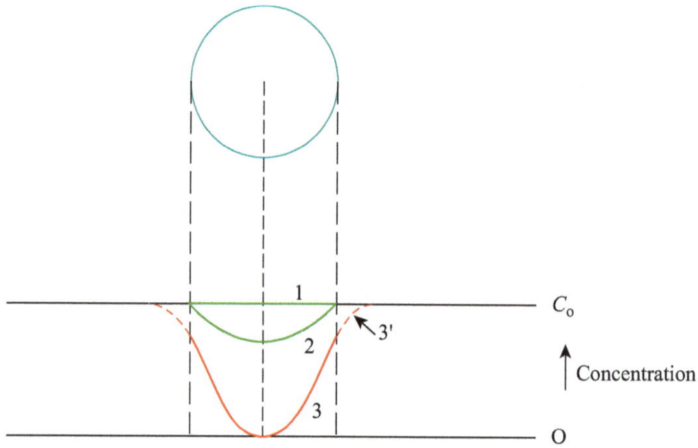

Figure 7.8:
Concentration profiles of a reacting molecule through a catalyst particle as a function of reaction conditions; 1, low temperature, large pores; 2, higher temperature, smaller pores; 3, high temperature, small pores.

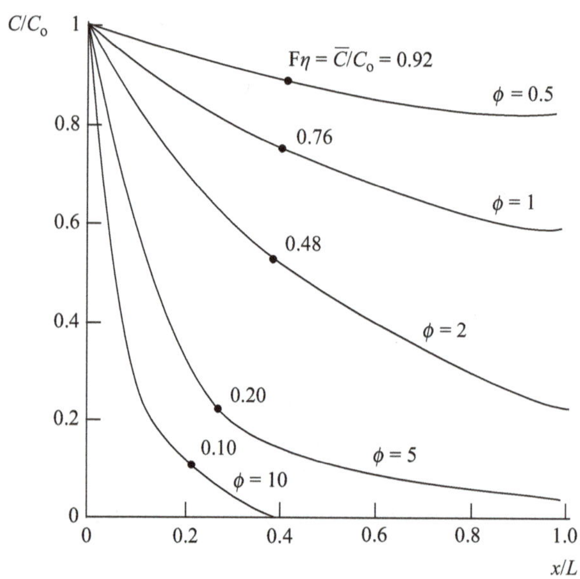

Figure 7.9:
Concentration profiles in a catalyst particle as a function of the Thiele modulus ϕ. The average internal concentration is also shown for each curve as a value of C/C_o.

(plotted as X/L) for different values of the Thiele modulus ϕ. (For a value of $\phi < 0.3$, not shown, the line would be parallel to the X/L axis.) Also shown are the average concentrations of the reactant divided by the concentration at the pore mouth, in other words, the effectiveness factor, η. It can be seen that concentrations at the centre of the pore are in all cases significantly lower than the concentration at the pore mouth and that the effectiveness factors can drop below 10% at high values of ϕ.

As discussed above, the Thiele modulus, ϕ, a dimensionless quantity, is a measure of the rate of the catalytic reaction relative to the rate of diffusion of the reactants in the pores of the catalyst (internal mass transfer); low values correspond to situations with high rates of internal mass transfer and high values to situations where internal mass transfer determines the effective rate of reaction. In order to ensure that mass transfer is at a minimum and to allow full use of high catalytic activity, it is therefore necessary to have active surface areas that are as accessible as possible. This can be achieved by using particle geometries that aid accessibility of the reactants to the active surfaces by having wide, short pores. One way to ensure this is to use very small particles that can be used in fluidized beds (see below). Another solution, as discussed previously, is to use support shapes designed to give external areas that are as high as possible while having relatively short pores; examples of suitable geometries – spherical particles, Raschig rings, hollow extrudates or tri-lobes – were shown in Figure 7.1. It is also possible to use geometries in which the catalyst is applied in relatively thin layers to an inert matrix, as in a monolith material or in micro-channelled reactors in which the micro-channels are coated with catalytically active components; see Chapter 5, Section 10.

7.3.3 Effects of Diffusion Limitations on Selectivity

In addition to having a hand in determining the activity of a catalyst, the effectiveness factor, η, can play an important role in relation to its selectivity. Consider a *sequential reaction*:

$$A \rightarrow B \rightarrow C \tag{7.6}$$

in which B is the desired product and C is an undesired by-product. (An example would be a selective oxidation reaction in which the selective product can be further oxidized to give CO_2 and H_2O.) The active sites for the reaction are located evenly through the pores of the catalyst and both steps in the reaction can occur on these sites; A and B will compete for the sites and so further reaction of B to give C only occurs if the concentration of A is low. (For simplicity, we assume that the pores are all cylindrical and open-ended, as in the case for Figure 7.9.) When η is unity, the concentration of reactant remains the same throughout the pores (see Figure 5.10) and so all of the active sites present in the pores of the catalyst contribute equally to the conversion of the molecule A to give B; hence, the selectivity to B is optimal. However, if η is low, for example at higher temperatures, the concentration of the reactant A at the centre of the pore drops to zero and the molecule B can diffuse more deeply into the pores to react further at empty sites. Hence, the selectivity of the catalyst now falls below its optimal value. To prevent such a situation occurring, it is possible to make so-called *egg-shell catalysts* in which the active phase is concentrated near the pore mouths (see Box 7.5). In such a geometry, even if B diffuses into the centre of the catalyst, there are no sites at which the production of C can occur. The use of egg-shell catalysts favours higher selectivities in sequential reactions and enables the use of higher reaction temperatures than would be possible using a catalyst made in the normal way.

BOX 7.5 Egg-shell Catalysts

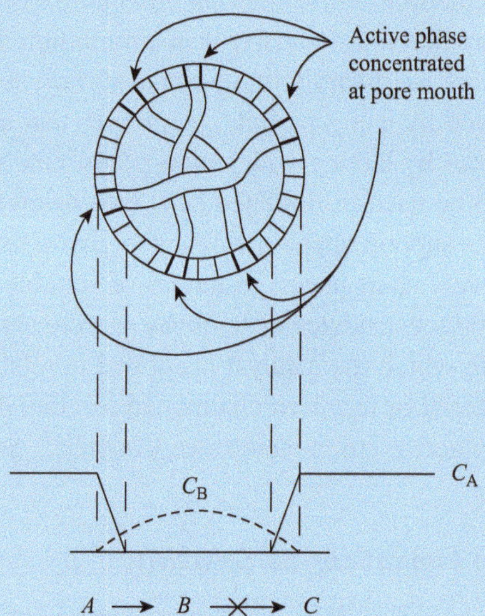

An egg-shell catalyst will be beneficial for a sequential reaction such as

$$A \rightarrow B \rightarrow C$$

if the reactions of A and B occur on the same site and full reaction of A occurs within the catalyst layer, as shown. An alternative for such catalysts is to avoid completely diffusional limitations, for example by using small particles with large pores, possibly in a fluidized bed reactor.

If a catalyst can bring about two different *competing reactions*:

$$A \rightarrow B; \text{ rate constant } k_A \tag{7.7}$$

$$X \rightarrow Y; \text{ rate constant } k_X \tag{7.8}$$

without diffusion limitations, the selectivity is given by:

$$S = k_A/k_X \tag{7.9}$$

However, with diffusion limitations:

$$S' = \sqrt{k_A D_A / k_X D_X} \tag{7.10}$$

where D_A and D_D are the respective diffusion coefficients. As the diffusion coefficients of *A* and *X* are unlikely to be very different,

$$S/S' \cong \sqrt{k_A / k_X} \tag{7.11}$$

For a selective reaction, it is probable that $k_A >> k_X$. Hence, we can conclude that the catalyst will be less selective when there are diffusion limitations than when there are none. Only if $k_X > k_A$ would there be an improvement in selectivity.

If there are two *parallel reactions*:

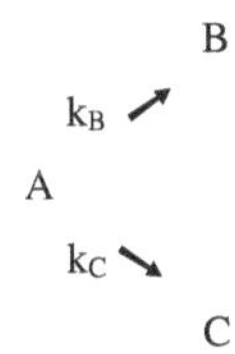

diffusion will not affect the selectivity unless the reactions have different orders; if the orders are different, the higher order reaction will be slowed more than the lower order reaction.

7.3.4 Effects of Diffusion Limitations on Catalyst Poisoning

If a reaction proceeding under diffusion limitations is poisoned by some component present in the reaction mixture by the selective adsorption of that component on the active sites, the sites at the pore mouth will become selectively poisoned (Figure 7.10), leaving the interior of the catalyst pores unpoisoned. A small amount of poison will therefore have a major effect on the catalyst activity. (A. Wheeler, Adv. Catal. 3 (1951) 349; see also Section 9.7 of J.J. Carberry's *Chemical and Catalytic Reaction Engineering*, Courier Dover Publications, 2001; http://books.google.com/books?id=arJLaKa4yDQC)

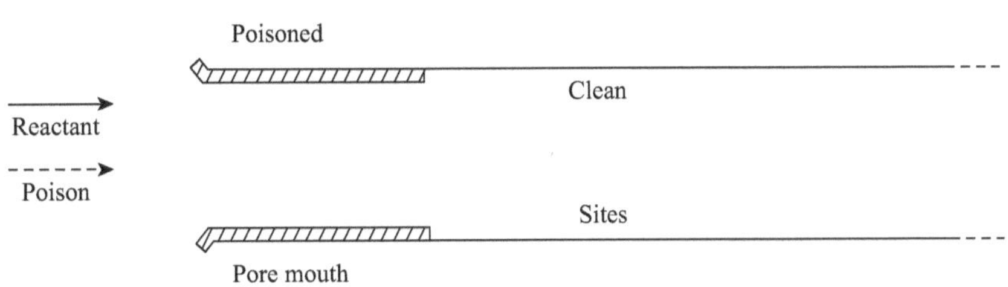

Figure 7.10:
Effect of diffusion limitations on catalyst poisoning.

7.4 Heat Transfer in Catalysis

Just as mass transfer (external and internal diffusion) plays an important role in practical applications of catalysis, so does heat transfer. With few exceptions, catalytic processes are either exothermic or endothermic and so the removal or supply of heat is very important. Further, as many of these processes are thermodynamically reversible, the uptake or evolution of heat can have a significant effect on the equilibrium position. Before discussing the design of some typical process plants, we will first consider the importance of temperature control on catalysed reactions.

7.4.1 Sequential Reactions

We will initially examine the effect of temperature on the course of a hypothetical reaction (shown in Eq. 7.6) that occurs in two steps:

$$A \xrightarrow{k_A} B, \text{ Activation energy, } E_A \tag{7.6a}$$

and

$$B \xrightarrow{k_A} C, \text{ Activation energy, } E_B \tag{7.6b}$$

We first answer the question at what temperature does one need to operate to obtain the maximum yield of the product B. Figure 7.11 gives two possible Arrhenius plots for these two reactions, one showing the case for $E_A > E_B$ and the other that for $E_B > E_A$. In the first case (Figure 7.11(a)), the rate constant for the formation of *B* from *A* will be much greater than that of C from B as long as the temperature is sufficiently high; that is, to the left-hand

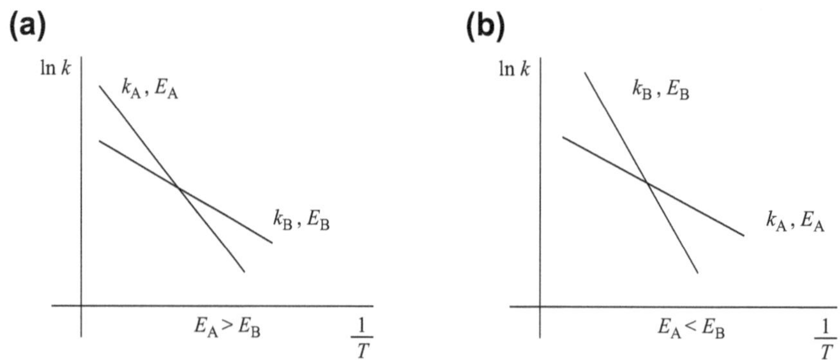

Figure 7.11:
Arrhenius plots for two (consecutive) reactions: $A \rightarrow B$ and $B \rightarrow C$ with (a) $E_A > E_B$ and (b) $E_B > E_A$.

side of the point of intersection. This means that the rate of the reaction from A to B will be much higher than the rate to form C, particularly when the concentration of B is low. On the other hand, for the second case (Figure 7.11(b)), the rate constant and hence the rate of the reaction of A to give B will only be significantly higher than that of B to give C if the temperature is low or if the concentration of B is low. Hence, to maintain a good selectivity to B, the temperature must remain well below that corresponding to the point of intersection in Figure 7.11(b). The drawback of such an approach in practice is that the rate of conversion of A will be relatively low. Hence, in order to decrease the time taken to obtain B, a different approach is taken: optimizing the time at any temperature in order to optimize the yield of B without the formation of any significant amount of C. This can be achieved by starting the reaction at high temperature and then going in steps to lower temperatures as shown very schematically in Figure 7.12. In this figure, which plots the concentrations versus the time in the reactor (contact time), results are shown for three decreasing temperatures, $T_1 > T_2 > T_3$. During operation at temperature T_1, k_B is much greater than k_A but, because the concentration of B is still quite small, the rate of formation of C is also small. As soon as an appreciable concentration of B has formed, C also begins to form and the rate of that reaction increases as B increases further. At that point, the temperature is decreased, lowering both the rates. Because E_B is greater than E_A, the rate of formation of C relative to that of B will be lowered and the latter reaction predominates. The same applies when the temperature is decreased once more.[4] Hence, remembering that the time of reaction is directly related to the catalyst bed length, the situation depicted in Figure 7.12 corresponds to three sequential reactors, each at a different temperature descending in the order T_1 to T_2 to T_3. The situation that we have just dealt with might correspond to a selective oxidation reaction in which B is the selective oxidation product and C is the non-selective product,

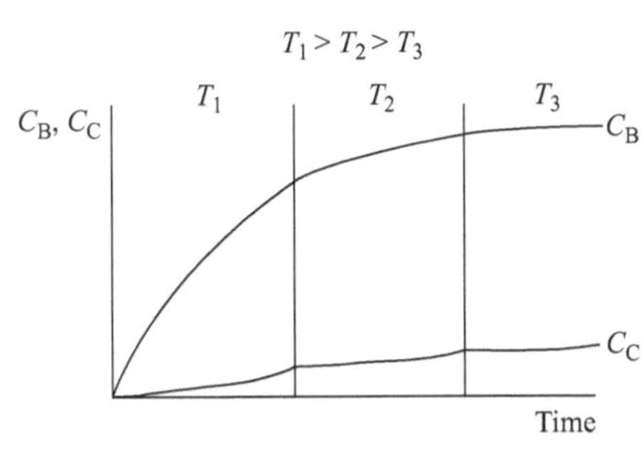

Figure 7.12:
Plots of C_B and C_C at three decreasing temperatures as a function of contact time for case (b) of Figure 7.11. This example assumes isothermal operation.

4 If the whole process had been carried out at the lowest temperature, the selectivity to B would have been higher than that achieved in the step-wise process. However, the contact time required to achieve that result would have been much greater.

CO_2. In that case, there would of course be other reactants and products but the arguments are the same.

7.4.2 Reversible Reactions

These arguments also apply equally to reversible reactions. Let us consider the reaction:

$$A \underset{k_B}{\overset{k_A}{\rightleftarrows}} B \tag{7.12}$$

This is entirely equivalent to the case of Eq. (7.6) in which the second reaction is the reverse of the first. Case (a) above (see Figure 7.11) then corresponds to an endothermic reaction for which

$$\Delta H^o = E_A - E_B > 0$$

while case (b) corresponds to an exothermic reaction.

$$\Delta H^o = E_A - E_B < 0$$

In order to get the optimal yield in case (a), as in the treatment above for sequential reactions, the endothermic forward reaction must be carried out at as high a temperature as possible in order to get optimum yields. An example of such behaviour is the steam reforming of methane, a reaction which we will discuss in detail elsewhere. In that case, the temperature of reaction is about 900 °C and this gives conversions of about 99%. The upper temperature in this case is limited by the availability of suitable reactor construction materials. As discussed in Chapter 8, the energy required to drive the reaction is provided by the combustion of natural gas. The reaction is carried out in long tubular reactors filled with suitable catalyst shapes, normally Raschig rings, and the tubes are heated externally along their full length by a furnace in which the natural gas is combusted.

In the case of an exothermic reaction, case (b), a different approach has to be followed. Figure 7.13 shows the temperature profile as a function of bed position (*or* time in the reactor *or* *W/F*, as long as all the catalyst beds have the same composition) through a series of four adiabatic catalyst beds for a typical endothermic reaction.[5] The entrance to the first bed is at a temperature of 400 °C, high enough for the catalyst to have sufficient activity to give the required rates of reaction. Immediately the reaction begins at the entrance of the bed, heat is given out and the temperature of the bed increases; as long as there is no significant axial heat

[5] Adiabatic operation assumes that there is no loss of heat from the reactor and that any heat generated is used to heat the catalyst in the immediate vicinity of the reaction. The example shown in Figures 7.13 and 7.14 corresponds to the oxidation of SO_2 to SO_3 but it could equally well apply to the reactions such as the synthesis of ammonia, the methanation of CO or methanol synthesis from syngas. The catalyst for the SO_2 oxidation reaction is a molten salt of K_2S_2O containing dissolved V_2O_5 retained in the pores of a porous silica support.

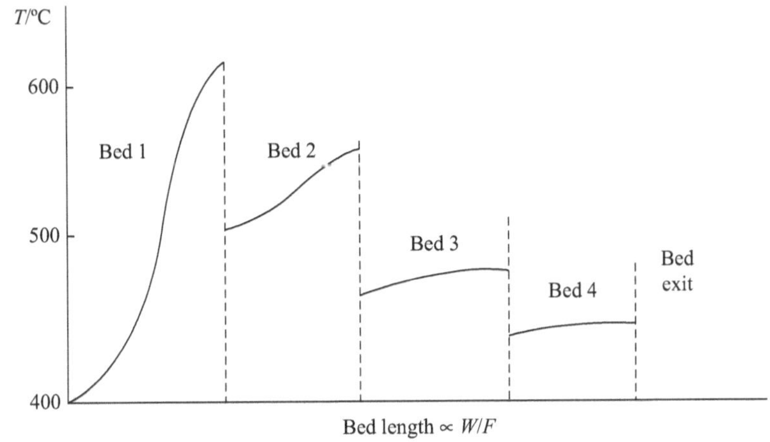

Figure 7.13:
Temperature profiles in a series of adiabatic reactor beds for a reversible exothermic reaction. To achieve this profile, there has to be cooling between each bed. (Taken from the lecture notes referred to in footnote 2.)

transfer and no heat is lost from the walls,[6] the temperature rise at any position in the bed is proportional to the conversion level at that point. The bed temperature then rises significantly until (see further below) the conversion is approaching the equilibrium value at the highest temperature. As in the example of the consecutive reactions dealt with above, the reaction mixture, including the products, is then passed to a second bed, the entrance of which is at a lower temperature. The all-over rate of the forward reaction is now much lower than at the same temperature in bed 1 since the rate of the reverse reaction has now become significant; nevertheless, bed 2 brings about further reaction, the bed length being designed such that the conversion at the exit of this bed approaches, but does not reach, the thermodynamic equilibrium position. The same applies in two further beds.

Figure 7.14 shows the data of Figure 7.13 for the four beds replotted as the values of the reactor temperature as a function of the conversion. The gradient of each line is the same, being determined by the heat capacity of the beds, each assumed to have the same construction and composition. (As above, it is assumed that the reactor system operates adiabatically. There will be significant deviations for non-ideal operation.) Also shown in Figure 7.14 are two curves, one corresponding to the equilibrium position for any temperature ("equilibrium line") and the another corresponding to the temperature giving the highest rate of reaction at any conversion ("optimal temperature line"). As shown in Box 7.6, the optimal temperature line can be calculated from a detailed knowledge of the kinetics of the reaction for the reversible reaction of *A* to give *B*.

[6] This is a somewhat idealized picture as there will inevitably be loss of heat from the reactor. This approach also assumes that the heat capacity of the bed does not change appreciably with temperature.

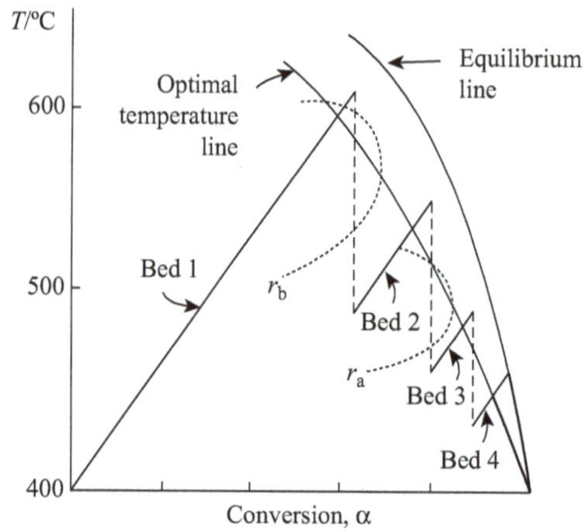

Figure 7.14:
The operating lines for a series of connected reactors for a reversible exothermic reaction. The solid curves represent the equilibrium conversions and the optimal temperature line. The dotted curves r_a and r_b represent lines of constant reaction velocity (see Box 7.6).

BOX 7.6 Derivation of the Equation Governing the Optimal Temperature Line for $A \rightleftharpoons B$

The rate of the forward reaction is given by:

$$r = r_A - r_B = k_A[A] - k_B[B] = k_A^o \cdot e^{-\frac{E_A}{RT}} \cdot [A] - k_B^o \cdot e^{-\frac{E_B}{RT}} \cdot [B]$$

The maximum rate of reaction occurs when $dr/dT = 0$. (Two dotted lines in Figure 7.14, r_a and r_b, represent constant rates of reaction; the tangent to these lines is parallel to the T axis at the optimal temperature line.)

$$\frac{dr}{dt} = 0 = \frac{1}{RT^2}\{k_A \cdot E_A \cdot [A] - k_B \cdot E_B \cdot [B]\}$$

$$= \frac{1}{RT^2}\{r_A \cdot E_A - r_B \cdot E_B\}$$

$$\therefore r_A \cdot E_A = r_B E_B$$

or

$$\frac{r_A}{r_B} = \frac{E_B}{E_A} = \frac{k_A^o}{k_B^o} \cdot e^{-\frac{(E_A - E_B)}{RT}} \cdot \frac{[B]}{[A]}$$

$$= \frac{k_A^o}{k_B^o} \cdot e^{-\frac{(E_A - E_B)}{RT}} \cdot \frac{1-\alpha}{\alpha} = 1$$

Hence, the corresponding value of α is given by:

$$\alpha/(1-\alpha) = k_A^o/k_B^o \cdot e^{-(E_A - E_B/RT)}$$

Similar expressions can be worked out for more complex kinetics.

7.4.3 Reactors for Exothermic Reversible Reactions

Just as for the sequential reaction $A \rightarrow B \rightarrow C$, the optimal yields for a reversible reaction are often achieved in practice in a sequence of reactors in which the reaction is started at as high a temperature as possible and then cooled down before admission to a second reactor. Figure 7.15 shows schematically the construction of such a reactor in which inter-bed cooling is achieved by the introduction of a coolant between the beds. (For simplicity, the jacket surrounding the whole of the reactor is omitted from the drawing.) In the geometry shown, frequently applied for the production of ammonia but also to the production of methanol from syngas and of sulfuric acid by the oxidation of SO_2 to SO_3, all exothermic reversible reactions, the feed is first passed through the "feed-effluent heat exchanger" in which it is heated by the exiting product gas. It then passes the outside shells of each of the three reactors, absorbing some of the heat generated in each of these reactors before passing to the entrance to bed 1. As the reaction proceeds, heat is evolved and the bed heats up. The gas leaving bed 1 is then cooled by heat exchanger pipes before entering bed 2, and so forth. The

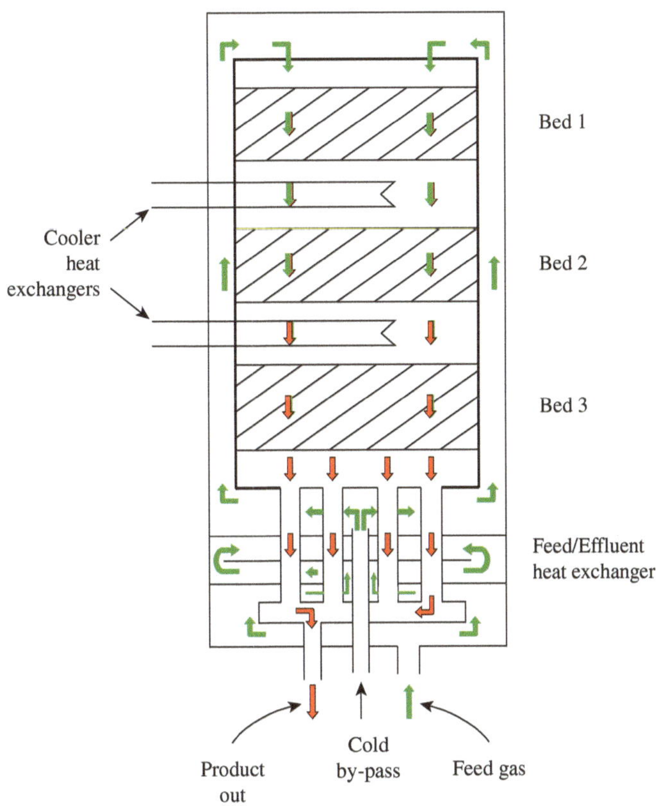

Figure 7.15:
Schematic representation of an adiabatic reactor used for the synthesis of ammonia. (The green arrows represent the feed and the red arrows represent the product.)

progress of the reaction in each bed will be given by plots similar in form to those shown in Figures. 7.13 and 7.14.

One of the problems with the geometry shown in Figure 7.15 is that each reactor has to be packed with catalyst separately and the removal and replenishment of catalyst charges can be quite complicated and time consuming. As a consequence, single bed reactors have been developed in which part of the reactant gas is introduced into the bed at various positions as a quench gas through nozzles suitably placed in the bed, this giving rise to similar temperature profiles to those discussed above; each addition of quench gas gives a decrease in temperature as in the reactor with inter-bed cooling discussed above. Figure 7.16 shows such a quench reactor developed by ICI (now Johnston Matthey) for ammonia synthesis in which the quench gas is introduced through nozzles placed at two positions in a single bed in such a way that there are again effectively three distinct beds; because of the simplified construction, the catalyst recharging process can be carried out relatively easily by removal from the base of the bed and addition of a new charge from the top.

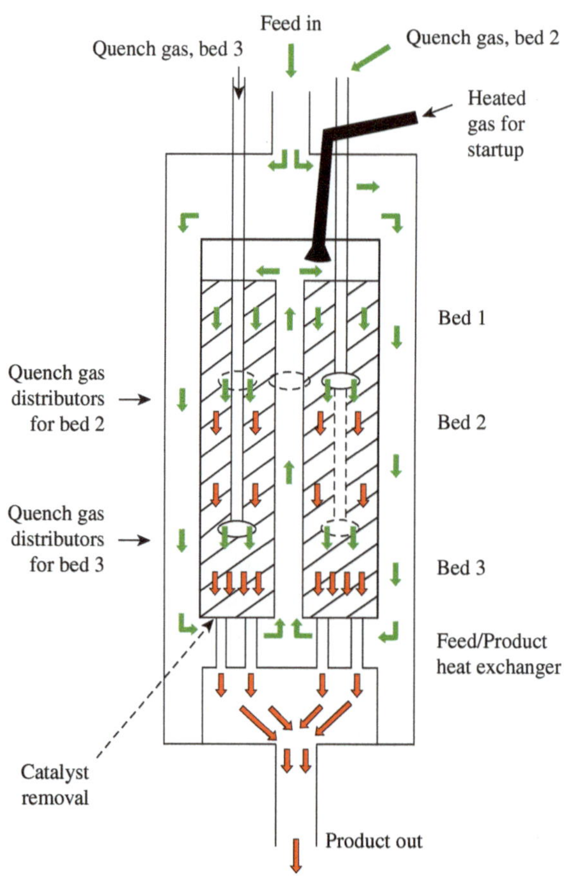

Figure 7.16:
A quench reactor developed by ICI for ammonia synthesis. (The green arrows represent the feed and the red arrows represent the product.)

A rather different approach is shown in Figure 7.17. This diagram depicts a "Radial Flow Reactor" developed by Haldor Topsøe for use in ammonia synthesis. Two separate beds are used but the flow is now arranged to pass radially through the bed, this having the advantage that diffusion lengths through the catalyst beds are kept to a minimum, thus minimizing pressure drops through the beds. Such an approach can be used to allow the use of smaller catalyst particles in order to decrease internal mass transfer limitations. After passing through the usual sequence of heat exchangers, the feed gas passes up through the centre of the second bed and diffuses through the wall of the inner tube of the first bed. The product gas then travels down the outer wall of the second bed through which it diffuses. The product leaving the second bed then heats the feed gas entering the first bed. Cooler gas can be added at the base of the centre feed tube in order to provide temperature control. It is claimed by Haldor Topsøe that this geometry gives very significant gains in the available catalyst surface area per unit reactor volume.

All catalytic processes require careful heat management and these examples have been given to illustrate the complexity of the design of the plants involved. As far as the catalytic chemist is

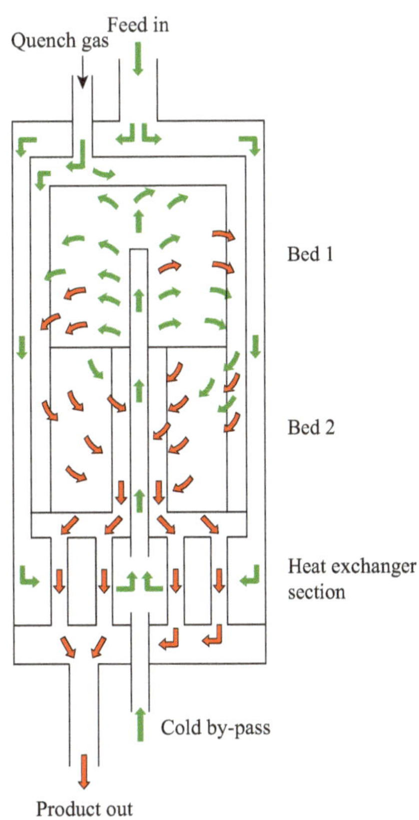

Figure 7.17:
A Haldor Topsøe Radial Flow Reactor for ammonia synthesis. (The green arrows represent the feed and the red arrows represent the product.)

concerned, the challenges of such types of application are many. The catalysts used must have the desired activity and selectivity, they must also have suitable geometries to enable maximal use of the active sites (in other words, minimizing diffusion limitations as discussed above) and they must have mechanical and thermal strength to withstand use under extreme temperatures in a practical reactor. An example of a process in which heat management is of paramount importance and in which the demands on the catalysts used is discussed in the following concluding section.

7.4.4 The Chemical Heat Pipe: "The Adam and Eve Process"

The so-called Adam and Eve Process was a concept pioneered in Germany in the 1970s at the Kernforschungsanlage (KFA) Jülich. Under development in the KFA at that time was a high temperature nuclear reactor in which the spherical fuel elements were cooled by helium gas, the helium exiting the reactor being at a temperature of 800 °C or above. The EVA reactor made use of this high-temperature helium to provide the heat necessary for the steam reforming of methane to produce syngas:

$$CH_4 + H_2O \rightarrow CO + 3H_2 \tag{7.13}$$

Whereas in the normal steam reforming reactor, the reforming tubes are placed in a gas-fired furnace and the tube walls are heated directly by radiant heat, the EVA reactor used a counter-current flow system in which there was direct heat transfer through the reactor walls from the helium to the reactor. As helium has exceptionally good heat transfer properties, the efficiencies of this reactor were very high.[7,8]

Having carried out the reforming reaction at the site of the nuclear reactor, the syngas produced was to be transported by pipeline to some distance (50–100 km) from the nuclear plant to a region where the energy stored in the syngas was required. The energy was then reclaimed by carrying out the reverse methanation reaction in the ADAM reactor:

$$CO + 3H_2 \rightarrow CH_4 + H_2O \tag{7.14}$$

After condensing out the water formed, the methane was then transported back to the nuclear reactor using a parallel pipeline and the cycle was completed. A schematic representation of this process is given in Figure 7.18 and a more detailed diagram of the reactor set-up is shown in Figure 7.19.

[7] Some preliminary work with this system is described in a paper by B. Höhlein, R. Menzer, J. Range, Appl. Catal. 1 (1981) 130.

[8] The Adam and Eve concept was tested in two pilot plants of differing scale. Because of the difficulty of coupling the system to the nuclear reactor, both these pilot plants used an electrically heated helium stream. Although both plants operated well, the concept was never taken to practical application as the high temperature nuclear reactor development work was abandoned.

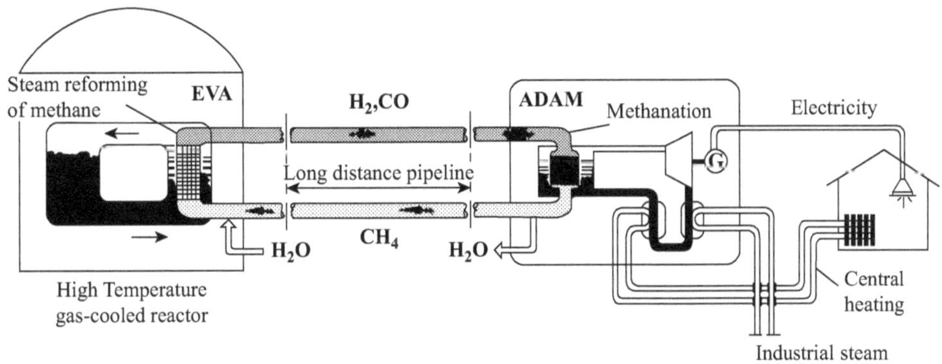

Figure 7.18:
Schematic representation of the ADAM and EVE concept. *Reproduced from the paper of footnote 7 with kind permission of Elsevier.*

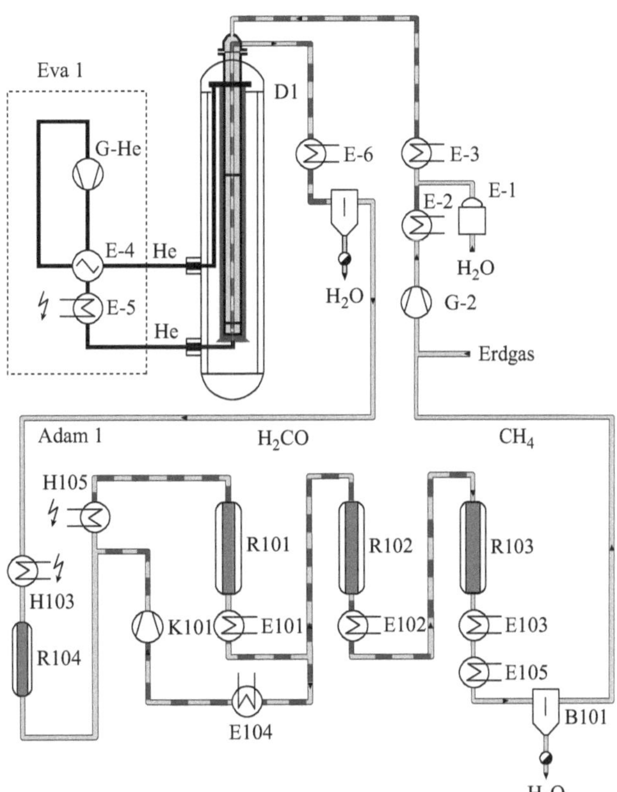

Figure 7.19:
The Adam and Eve system, showing details of the reactors, feed systems, etc. *Reproduced from the paper of footnote 7 with kind permission of Elsevier.*

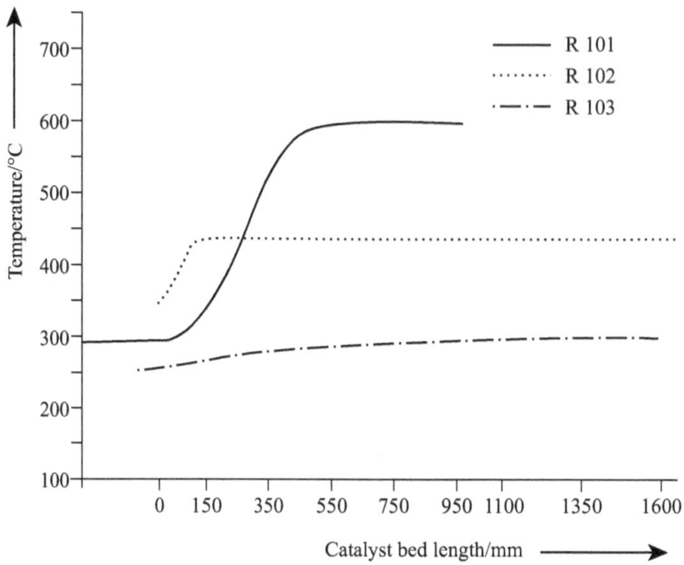

Figure 7.20:
Temperature profiles in the ADAM 1 reactor. *Reproduced from the paper of footnote 7 with kind permission of Elsevier.*

In the context of this chapter, the geometry of the ADAM system is particularly informative. In order to achieve good conversions back to methane, a sequence of three reactors with inter-bed cooling of the type described above for the ammonia synthesis reaction was used. However, because the enthalpy of the methanation reaction is so high, the temperature rise in the first bed would be about 800 °C if an undiluted feed was used with an entrance temperature of about 300 °C. In order to prevent such a large temperature increase, a recycle loop was introduced and this had the effect of decreasing the upper temperature to about 600 °C. At the time when this work was initiated, the only methanation catalysts available commercially were those used for the methanation of traces of CO present in the hydrogen stream of ammonia prior to entering the ammonia synthesis reactor. The temperature rise in such a methanation reactor is small and so the demands on the catalyst are relatively straightforward: sufficient activity at about 450 °C with adequate thermal stability at that temperature to ensure that the catalyst does not deactivate during use over extended periods. The requirements for the Adam reactors were different: adequate activity at 300 °C to start the reaction and sufficient thermal stability at 600 °C to resist the severe sintering that would occur with standard methanation catalysts. The work led to the development, in collaboration with Haldor Topsøe, of new Ni-based methanation catalysts which exhibited the required activity and stability. Typical results for the temperature profiles for the three beds of Adam 1 using a Haldor Topsøe catalyst are shown in Figure 7.20. The larger Adam and Eve II pilot plant operated for some time in the early 1980s, producing sufficient heat for the needs of a small town. However, political pressures led to the abandonment of the high temperature reactor and hence of this process concept.[9]

TASK 7.2

The idea behind the Adam and Eve system has more recently been adopted in work aiming to harness sunlight using CO_2 reforming of methane as the endothermic reaction. The reforming reactor is placed at the focus of the heliostat and the syngas produced is piped to where the energy is needed. Examine the literature to learn something more about the use of such systems in solar applications using the paper by M. Levy et al., Solar Energy 50 (1993) 179–189 as a starting point. See if you can find out about other conceptual designs for chemical heat pipes, starting from the work of Zeng et al. using a system based on the reversible SO_2/SO_3 cycle. (Int. Atomic Energy Agency, Vienna (Austria), IAEA-TECDOC–761, pp. 102–107; http://www.iaea.org/inisnkm/nkm/aws/htgr/fulltext/25067248.pdf/)

[9] There is some evidence that pressure is currently mounting to recommence the development of this type of nuclear reactor.

CHAPTER 8

Some Catalytic Reactions

Chapter Outline

8.1 Introduction

In the previous chapters, you will have learnt much of the important background to catalysis and some of the important parameters which determine whether or not a catalyst can be used in a catalytic process. In this final chapter, we will describe briefly a variety of aspects of some important catalytic processes currently encountered in practice and will give pointers for further self-study using Scopus, Web of Science and other search tools. The emphasis – working on the assumption that you will learn much more by reading the original articles written by authors with hands-on experience than from a dry textbook – is on encouraging you to read the relevant literature. Wherever possible, you are given links to review articles which summarize the appropriate review author's knowledge of the field in question but occasionally

Heterogeneous Catalysis. DOI: 10.1016/B978-0-444-53363-0.10008-8

you will be pointed in the direction of a particularly important original paper. Do not try to read and understand each and every detail of anything you read – each time you read a paper you will find something different of interest; try instead to distil out the most relevant points relevant to your current search.

As has been emphasized in previous chapters, catalysis plays a pivotal role in almost all industrial processes. Indeed, it has been estimated that more than 80% of all commercial products, or some element of them, have been produced by catalysis. In the following sections, we will discuss the applications of catalysis in the production of energy carriers and bulk chemicals and will then briefly discuss the importance of the subject in fine chemicals and pharmaceuticals before ending with discussions of the applications of catalysis in environmental control and biomass conversion.

8.2 Catalysis in the Conversion of Natural Gas

Figure 8.1 shows schematically the importance of catalysis in the conversion of the natural resources, coal, oil and gas, to some useful products. The solid arrows 1, 2 and 3 represent the use of these resources for energy purposes, supplying heat and electricity by combustion processes,

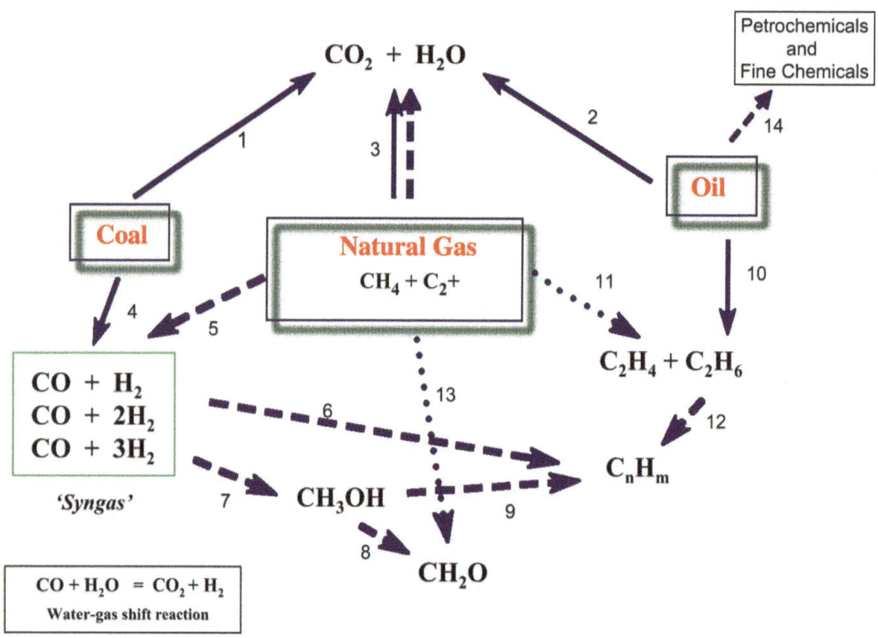

Figure 8.1:
Schematic formulation showing the production of energy and chemicals from coal, oil and natural gas. The solid arrows represent non-catalytic routes, the dashed arrows represent established catalytic processes and the dotted arrows represent possible future developments. (Route 14 represents the use of oil in the petrochemicals industry oil refineries and the petrochemicals and fine chemicals industries will be discussed in later sections.)

these processes being non-catalytic. The catalytic combustion of natural gas (see Box 8.1) is also shown for completeness as a dashed arrow (3); catalytic combustion has the advantage over homogeneous combustion that the emission of NO_X is minimized and it can also be used effectively to remove traces of methane or higher hydrocarbons from other gases. The only other non-catalytic processes shown are coal gasification (solid arrow 4) and the thermal cracking of oil fractions to give olefins (solid arrow 10). For completeness, the conversion of oil fractions to petrochemicals and fine chemicals by catalytic means is shown by the dashed arrow (14); we will cover both of these aspects of the application of catalysis separately in Section 8.3. We will now discuss the other conversion processes in turn. (We have already encountered a number of the other processes in preceding chapters; cross-references are given as appropriate.)

BOX 8.1 Catalytic Combustion of Methane

One of the most commonly used catalyst for methane combustion is Pd supported on alumina or some compound containing alumina such as mullite or cordierite. The following figure (reproduced with modification from J.H. Lee, D.L. Trimm, Fuel Process. Technol. 42 (1995) 339–359) shows several features of the reaction occurring over a typical catalyst formulation.

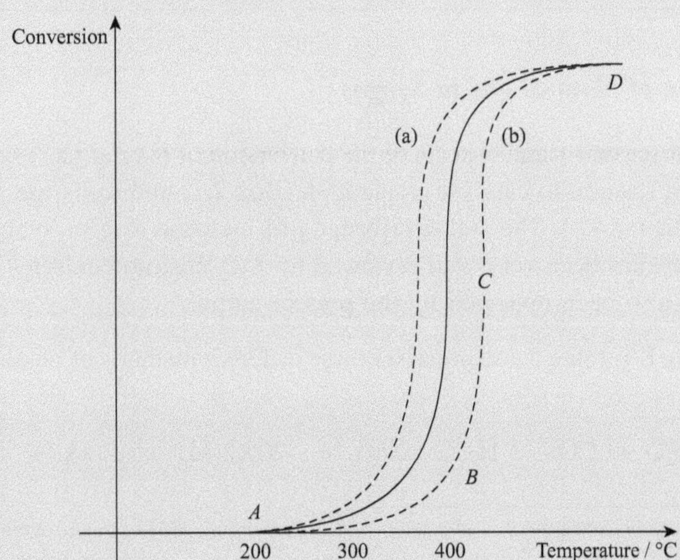

The solid curve shows schematically typical behaviour observed over a Pd/alumina catalyst, the conversion being plotted as a function of the temperature. Region *A* shows the typical "light-off" behaviour of such a catalyst and region *B* represents results obtained under kinetic conditions. However, as the reaction itself is highly exothermic, it is difficult to control the temperature and this rises so that the behaviour (region C) is typical of mass transfer control (internal diffusion

(*Continued*)

BOX 8.1 Catalytic Combustion of Methane—*Cont'd*

limitations). Finally, in region *D*, at ca. 100% conversion, the reaction becomes controlled by the rate of supply to the external surface ("reactant depletion"). The temperature scale shown is only approximate but it can be seen that light-off occurs around 200 °C while depletion occurs at ca. 400 °C. The upper temperature is well below the range in which the formation of NO_x occurs and is also below the range of temperature in which the gas-phase radical reactions are generally observed. There is a body of evidence to show that the active surface of the Pd catalyst is (at least partially) oxidized. The dotted curve (a) represents results obtained after the data for the freshly reduced catalyst (solid curve) had been pretreated in the oxygen-containing reaction mixture before the measurements, there being a significant improvement in the catalytic behaviour. The dotted curve (b) illustrates another aspect of all work with heterogeneous catalysts: the poisoning effect of various components of the gas feed used, in this case 100 ppm H_2S. Under the reaction conditions used, adsorbed sulfate species are formed which can then be removed once more by thermal treatment under nitrogen or under vacuum.

The review by Lee and Trimm referred to above gives an excellent description of some of the work reported in the literature on methane combustion; the effects of activation and of H_2S addition shown in the figure above are from results originally reported by L.J. Hoyos, H. Praliaud, M. Primet (Appl. Catal. A 98 (1995) 125—138).

8.2.1 Conversion of Natural gas to Syngas

We have already discussed some aspects of the conversion of natural gas to syngas (Reaction 4 of Figure 8.1) in relation to catalyst preparation (Box 7.1) and to its use in the Adam and Eve process (Section 7.4.4). The steam reforming of methane to give syngas is a mature technology and this has been very well reviewed by J.R. Rostrup Nielsen.[1] Some aspects of the reaction have also been reviewed by the present author.[2]

As shown in Figure 8.1, route 5 encompasses three different methods of obtaining syngas from methane:

$$CH_4 + H_2O \rightarrow CO + 3\ H_2 \qquad \Delta H^\circ = -206.2\ \text{kJ} \qquad (8.1)\quad \text{Steam reforming}$$

[1] Rostrup Nielsen has published a number of papers on steam reforming. One of the most comprehensive reviews on the subject, entitled "Catalytic Steam Reforming", is published in *Catalysis – Science and Technology*, Edited by J.R. Anderson and M. Boudart, Springer-Verlag, Berlin, Heidelberg, New York and Tokyo, 5 (1984) 1—117. A more recent review describing recent developments is "Steam reforming and chemical recuperation", Catal. Today 145 (2009) 72—75. The interested reader is advised to carry out a comprehensive search through Scopus using the name of this author as a starting point.

[2] J.R.H. Ross, "The steam reforming of hydrocarbons", Surface and Defect Properties of Solids, Edited by M.W. Roberts and J.M. Thomas, Specialist Periodical Reports, Royal Society of Chemistry, London, 4 (1975) 34—67; "Metal catalysed methanation and steam reforming", Catal., G.C. Bond, G. Webb (Eds.), Specialist Periodical Reports, Royal Society of Chemistry, London, 7 (1985) 1—45; "Natural gas Reforming and CO_2 mitigation", Catal. Today, 100 (2005) 151—158.

$$CH_4 + 1/2\ O_2 \rightarrow CO + 2\ H_2 \qquad \Delta H^\circ = -36\ \text{kJ} \qquad (8.2)\ \text{Partial oxidation}^3$$

$$CH_4 + CO_2 \rightarrow 2\ CO + 2\ H_2 \qquad \Delta H^\circ = -247.4\ \text{kJ} \qquad (8.3)\ \text{“Dry” reforming}$$

All three reactions are shown as being irreversible but they are all in fact equilibria and can only be considered as being irreversible if carried out at high temperatures. Further, they are complicated by the fact that the water−gas shift reaction can also take place, altering in most cases the final gas composition (particularly in reactions (8.1) and (8.3) and, but to a lesser extent, in (8.2) if some total oxidation to give CO_2 also occurs):

$$CO + H_2O \rightleftarrows CO_2 + H_2 \qquad \Delta H^\circ = -41.2\ \text{kJ} \qquad (8.4)$$

One final and very important complication is the lay-down of carbon on the catalysts by two different reactions:

$$\text{Boudouard reaction } 2CO \rightleftarrows C + CO_2 \qquad \Delta H^\circ = -172.4\ \text{kJ} \qquad (8.5)$$

and

$$\text{Methane decomposition } CH_4 \rightleftarrows C + 2\ H_2 \qquad \Delta H^\circ = 74.9\ \text{kJ} \qquad (8.6)$$

Carbon deposition can be largely avoided (thermodynamically) in steam reforming (reaction 8.1) by operating in the presence of an excess of steam. However, it cannot be avoided thermodynamically with dry reforming (reaction 8.3) and so only kinetic control (developing a catalyst that does not favour carbon deposition) has any chance of success in dry reforming. In partial oxidation (reaction 8.2), the problem is less significant as long as the conditions are chosen such that any carbon formed can immediately be burnt off by even low partial pressures of oxygen in the feed. Although “dry reforming” gives an attractively low H_2/CO ratio in the products of approximately unity, the reaction is not currently practiced commercially, at least partially because of this carbon deposition problem.[4]

The following are some of the important aspects of the use of the steam reforming of methane:

- All systems must be preceded by sulfur removal using a hydrodesulfurization unit.
- Commonly used catalysts: nickel on refractory support.
- Reactor types: externally heated tubular reformers in which natural gas combustion supplies the energy needed.

[3] The partial oxidation reaction (8.2) occurs in two steps, the oxidation of part of the methane to CO_2 and H_2O followed by reactions (8.1) and (8.2) to consume the rest of the methane; oxidation occurs exothermically at the entrance of the catalyst bed and then the slower endothermic reactions occur further down the bed.

[4] Several processes such as the Midrex, Armco and Nippon Steel Purofer processes use the reaction to produce syngas for the reduction of iron ore; these processes recycle the gases exiting the ore reduction step and so the feed contains both CO_2 and water. (http://www.kobelco.co.jp/p108/dri/e/dri04.htm\) Haldor Topsøe have operated a dry reforming pilot plant using undiluted CO_2 feed but have not commercialized the technology.

BOX 8.2 Carbon Filament Growth

It is well established that the carbon which forms on catalysts by the carbon-forming reactions discussed here is in the form of filaments. Much research work has been carried out on the mechanism of filament formation and on the use of the filaments formed, for example as catalysts supports and adsorbents. Reaction (8.6) has also some potential as a source of CO-free hydrogen, even though the operation of such an approach would depend on there being a parallel economical use for the carbon formed. A recent review on this subject deserves attention: Yongdan Li*, Douxing Li, Gaowei Wang, "Methane decomposition to COx-free hydrogen and nano-carbon material on group 8–10 base metal catalysts: a review", Catal. Today 162 (2011) 1-48.

The current author has suggested that the formation of carbon during the CO_2 reforming reaction is an energetically favourable way sequestering CO_2 ("Natural gas reforming and CO_2 Mitigation", Catal. Today 100 (2005) 151–158).

- Pre-reformers: uses packed bed to convert any other components of the natural gas into methane.
- Product composition: close to equilibrium as determined by thermodynamic calculations.

Depending on the end uses, different strategies are used. For example, for hydrogen production in ammonia synthesis, water–gas shift reactors and a tail-gas methanation reactor are included. For syngas production for use in Fischer–Tropsch synthesis, autothermal reformers are added, these being fed with a mixture of steam and pure oxygen.[5]

CO_2 reforming of methane ("dry reforming") has been a "hot subject" for some years and continues to be so. The reaction and its catalysts have attracted a great deal of academic interest, despite the lack of industrial application. This is partly because of the misconception that the reaction can be used to remove unwanted CO_2. Even if catalysts could be found that would resist carbon deposition over the long periods necessary for industrial operation of the process, the potential worldwide use of any syngas formed would be extremely small in comparison with the amount of CO_2 formed annually from even a small number of electricity generating plants. Nevertheless, the search for effective catalysts has led to some very useful academic research. While nickel catalysts can generally be used for steam reforming applications, they generally do not resist carbon deposition for more than a relatively short periods of operation under CO_2 reforming conditions ($CH_4/CO_2 \approx 1.0$). On the other hand, a variety of noble metal catalysts appear to work reasonably well for extended periods, probably due to the fact that carbon can only form on the exterior of the metal crystallites, this being in contrast to

[5] P. Bakkerud, J. Gøl, K. Aasberg-Petersen, I. Dybkær, "Preferred synthesis gas production routes for GTL", Stud. Surf. Sci. Catal. 147 (2004) 13–18. Other papers in the same volume discuss in detail other aspects of GTL technology.

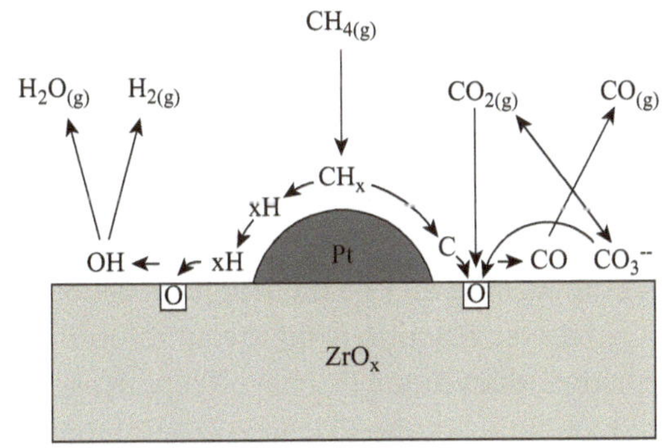

Figure 8.2:
A model for the CO_2 reforming of methane over a Pt/ZrO_2 catalyst. *Source: A.M. O'Connor, F.C. Meunier, J.R.H. Ross, Stud. Surf. Sci. Catal. 107 (1998) 819. Reproduced with kind permission of Elsevier.*

nickel which dissolves carbon in the bulk. Figure 8.2 (also shown in Box 3.6) shows a model of the situation existing on the surface of a Pt/ZrO_2 catalyst under dry reforming conditions. It appears that methane adsorbs on the surface of the Pt crystallites of the catalyst and dissociates there, liberating hydrogen; the carbon formed diffuses to the metal-support interface where it picks up an oxygen entity from the support. Finally, the support is re-oxidized by the CO_2 from the reactant mixture. The Pt surface remains largely covered by carbon but the catalyst retains its activity due to the sites at the metal-support interface remaining free of significant quantities of adsorbed carbon. It should be noted that the reaction scheme depicted in Figure 8.2 is a variant of the Mars-van Krevelen mechanism discussed in Chapter 6.

TASK 8.1 Catalysts for Dry Reforming

Some aspects of the CO_2 reforming of methane are discussed in the following article: J.R.H. Ross, "Natural gas reforming and CO_2 mitigation", Catal. Today, 100 (2005) 151–158. Using the references in this article together with citing articles found using Scopus, do an up-to-date survey of the catalysts which have been examined for use for dry reforming.

8.2.2 Fischer–Tropsch Synthesis

The syngas made by the processes discussed in the previous section has a number of uses, either as a source of pure hydrogen (e.g. for use in ammonia synthesis and hence the production of fertilizers, as discussed previously, or in hydrogenation processes) or as feed for reactions involving both CO and hydrogen. We discussed in the previous chapter the methanation reaction and its application in the Adam and Eve process. We will now discuss two other reactions in which syngas is used, Fischer–Tropsch synthesis and methanol synthesis (Reactions 6 and 7 of Figure 8.1 respectively). (The history of the development of both these processes was covered in Chapter 1.)

The Fischer–Tropsch synthesis of higher hydrocarbons

$$nCO + (n + m/2)H_2 \rightarrow C_nH_m + nH_2O \qquad \Delta H^\circ < 0 \tag{8.7}$$

uses the appropriate mixture of CO and hydrogen, making use of the water–gas shift reaction to alter the composition as appropriate. (As the predominant products are most frequently long-chain saturated paraffins, $m = 2n + 2$; hence, the required ratio is close to 2.0.) The reaction, strictly speaking, is reversible and so the single arrow could be replaced by a reversible one. (Indeed, the reverse reaction is the steam reforming of a hydrocarbon (rather than methane) and the thermodynamics of the Fischer–Tropsch and steam reforming processes are directly interrelated.[6]) High yields in the Fischer–Tropsch process will result if the temperature is low and the pressure is high (Le Chatelier's Principle). In practice, as with other reversible processes such as ammonia synthesis (discussed in Chapter 7), the reaction has to be carried out at sufficiently high temperatures to ensure reasonable conversions without restricting the conversion too much. Another important limitation of the process is that the product formed is almost always a mixture of molecules of very different molecular weights, ranging from methane to waxes, this being largely determined by a so-called Schultz-Flory distribution. This distribution arises from the mechanism of the reaction which involves a step-wise polymerization chain process in which the probability of chain growth versus termination is given by the symbol α.[7] Figure 8.3 shows a simplified plot of the weight fraction of any

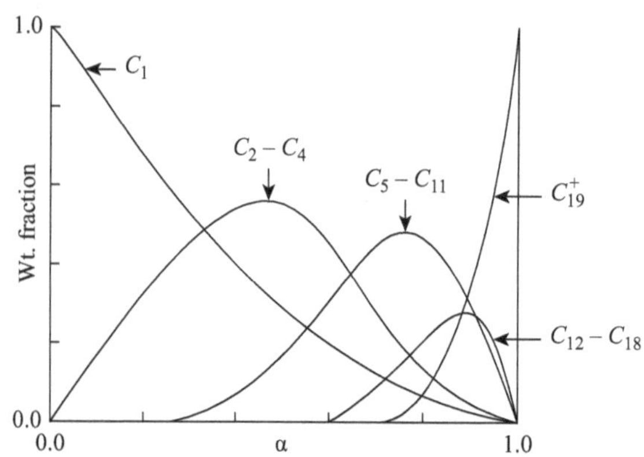

Figure 8.3:
Weight fraction of FT products as a function of α, the chain growth probability. *Reproduced with kind permission of Elsevier.*

[6] The present author reviewed various aspects of the steam reforming reaction, including its thermodynamics, in 1975 (J.R.H. Ross, "The steam reforming of hydrocarbons", Specialist Periodical Reports, Surface and Defect Properties of Solids, Eds. M.W. Roberts and J.M. Thomas, 4 (1975) 34–67).

[7] There are examples where the distribution deviates from the strict mathematical expression of the Schultz-Flory distribution. For an advanced treatment of this and other aspects of the Fischer–Tropsch Process, the reader should refer to more advanced texts such as *Fundamentals of Industrial Catalytic Processes*, C.H. Bartholomew, R.J. Farrauto, Wiley-Interscience, New Jersey, 2006.

particular product range to be expected for any value of α. It can be seen that with low α, methane is the expected product while at high α (approaching unity), only high molecular weight products are expected. In general, three different metallic components of catalysts have received attention: Ni, which has a low value of α and tends to give methane as a predominant product; Fe, which has an intermediate valuc of α and gives a relatively wide range of products; and Co which has a high value of α and tends to give high molecular products such as waxes.[8] Iron and cobalt are therefore the metals of choice.

BOX 8.3 Mechanism of the Fischer Tropsch Reaction

The original description of much of the early German work on the subject and a possible mechanism for the Fischer–Tropsch process was given by H.H. Storch, N. Gulumbic and R.B. Anderson in "The Fischer–Tropsch and Related Syntheses" (John Wiley & Sons, New York, 1951, 610 pp.). Their mechanism involved chain growth using oxygenated C_1 species ($CHOH_{(ads)}$). Strong support for this mechanism was given by P.H. Emmett and coworkers from their work using labelled molecules. The more recent, and more generally accepted, approach involves CH_2 species as the chain growth species. However, it seems that oxygenated species are still the predominant species if the main products of the reaction are alcohols.

Much has been written about the history of the development of the Fischer–Tropsch process. A very comprehensive compilation of the literature on the process has been assembled on a special web site: http://www.fischer-tropsch.org/. For the present purposes, it is sufficient to say that there have been four periods in the development of the process to its current position:

- Development of the original process and its use in Germany up until the end of the Second World War, when the majority of effort was devoted to Fe-based catalysts.
- The use of the process in South Africa during the period when that country was unable to import oil as a result of the trade embargoes in force and when SASOL operated a full-scale plant for the production of petroleum and diesel.
- Research in the US and elsewhere, particularly spurred on by the energy crisis of the early 1970s, when attention began to focus once more on Co catalysts.
- The modern development of gas to liquid plants (GTL) for use in situations where there is an excess of natural gas, generally associated with oil reserves and hitherto flared off; there have very recently also been developments in relation to biomass conversion to liquids and in relation to microchannel reactors.

8 The only other metal which has received any attention is ruthenium; however, it is very scarce and is therefore not usable as a catalysts component: any large-scale demand for its use would immediately give rise to a significant increase in price. Copper gives alcohol production, predominantly methanol, as is discussed in the next section.

TASK 8.2 Catalysts for Fischer Tropsch Synthesis

M.E. Dry was a senior scientist with Sasol during the period when the only significant work on Fischer–Tropsch synthesis occurred in South Africa. During the period of political isolation, the FT process was the only source of automobile fuel in South Africa. Some of the work at Sasol is described in the influential article: M.E. Dry, The Fischer–Tropsch Process: 1950–2000, Catal. Today 71(2002) 227–241. Analysis with Scopus shows that this article is still highly cited. Examine the citation behaviour, paying particular attention to work on new developments such as GTL and biomass conversion. Examine also the types of catalysts which have been studied and the effect of catalyst composition, reaction conditions, etc., on product distribution. You should also encounter reference to Schultz-Flory distributions of products. Include in your review of the literature an explanation of such distributions and their importance.

A particularly interesting review of the subject of the Fischer–Tropsch process and its catalysts, including some fascinating historical details only recently unearthed, has recently been provided by Casci, Lok and Shannon.[9] These authors have analysed the patent activity over the last 30 years and have demonstrated very clearly the very significant upsurge of interest over the last 10–15 years, particularly by Sasol, Shell and

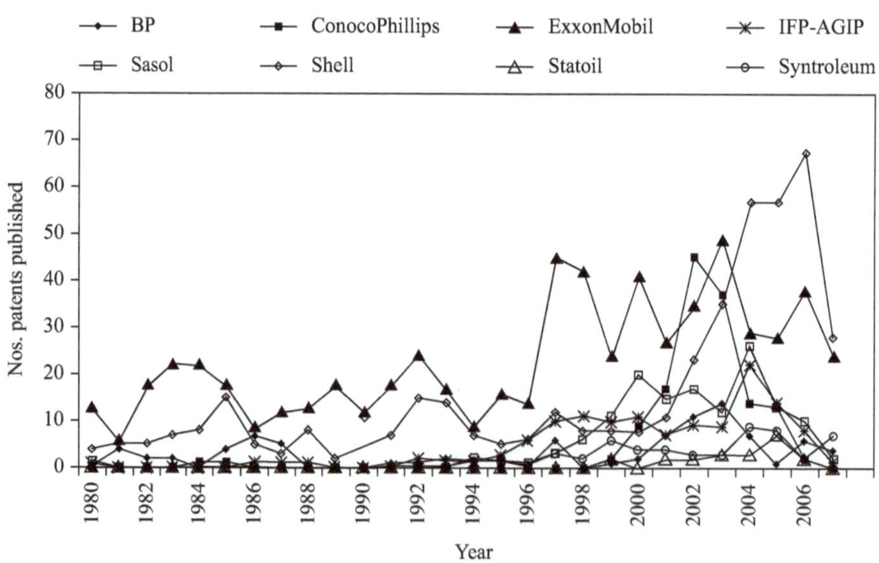

Figure 8.4:
Patents related to Fischer–Tropsch synthesis, with assignee, published since 1980.
Reproduced from Casci et al.[9] with kind permission of Elsevier.

[9] J.L. Casci, C.M. Lok, M.D. Shannon, "Fischer–Tropsch catalysis: the basis for an emerging industry with origins in the early 20th century", Catal. Today, 145 (2009) 38–44.

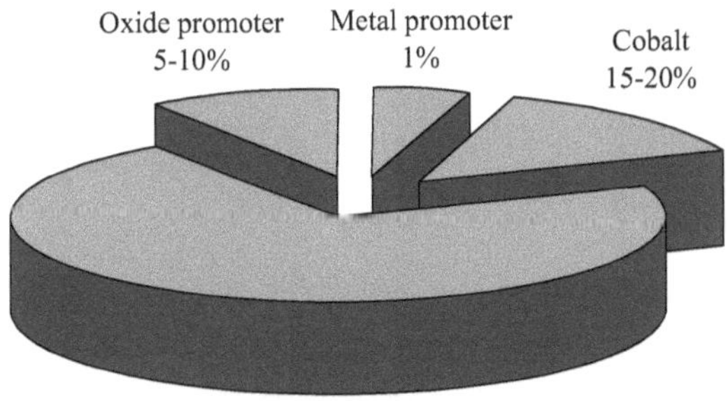

Figure 8.5:
Typical compositions of modern Co-based FT catalysts.[6] *Reproduced with kind permission of Elsevier.*

Exxon/Mobil; see Figure 8.4. They show that the majority of current effort concerns the development of Co-containing materials, often supported on preformed supports such as titania, silica or alumina using cobalt nitrate for the impregnation step; these catalysts often contain oxide and noble metal promoters; see Figure 8.5. As indicated above, most of the current development activity in the area of FT synthesis is directed towards the production of fuels in GTL plants. Several of these are being built and commissioned in the Gulf States although the first plant using the modern technology was built by Shell at Bintulu (Sarawak, Malaysia). Figure 8.6 shows a schematic representation of this plant. In essence, it first converts natural gas to syngas by a non-catalytic partial oxidation process. The syngas is then converted into paraffins by a Fischer–Tropsch (FT) process involving multi-tubular fixed bed reactors. Finally, the FT product is hydrocracked/hydro-isomerized to give a mixture of naphtha, middle distillates and a waxy residue. Further details of these developments and also some of the science behind them are available in a range of

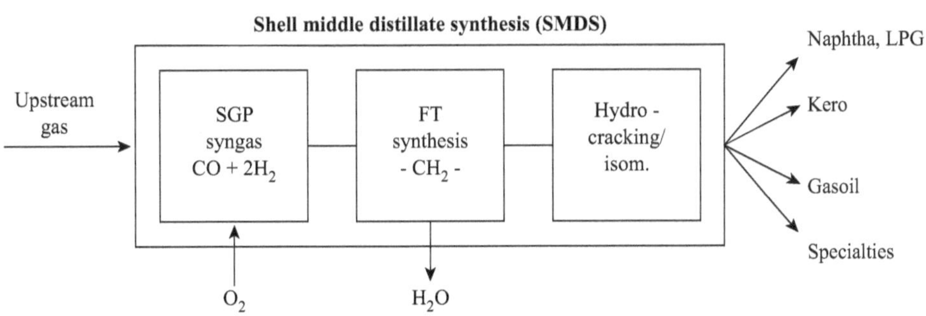

Figure 8.6:
The Shell middle distillates synthesis. *Source: A. Hoek, L.B.J.M. Kersten, Stud. Surf. Sci. Catal. 147 (2004) 25–30. Reproduced with kind permission of Elsevier.*

proceedings of the Natural Gas Conversion Symposia; see for example, Studies in Surface Science and Catalysis, vol.147 (2004), edited by Xinhe Bao and Yide Xu.

8.2.3 Methanol Synthesis and Methanol Reforming

The history of the development of the commercial synthesis of methanol (Figure 8.1, reaction 7) is in many ways similar to that of ammonia (see Chapter 1). Early work by Sabatier and Senderens in 1905 (P. Sabatier, J.B. Senderens, Ann. Chim. Plzys. 4 (1905) 418), aimed at developing a catalyst for methanol synthesis, showed that methanol decomposition could be brought about by a Cu catalyst but that the reverse (synthesis) reaction was less effectively catalysed. Thereafter, a patent by Patard (French Patent 540,343 (Aug. 1921)) claimed that a variety of materials, including ones containing Cu, were effective catalysts for methanol synthesis. However, despite this, the high pressure process that was subsequently developed by BASF in 1923 made use of a "zinc chromite" catalyst. This process which operated at a pressure of 200 atm and 350 °C was only superceded in the 1960s by the low pressure process pioneered by ICI (now Johnson Matthey). For a more detailed review of this work, see the extensive review: K. Klier, Methanol synthesis, Adv. Catal. 31 (1982) 243–313.

The catalysts developed by ICI were based on the $Cu/ZnO/Al_2O_3$ system. These catalysts bring about the effective conversion of syngas (also containing a proportion of CO_2) at 50–100 atm and 220–250 °C:

$$CO + 2\ H_2 \rightarrow CH_3OH \qquad \Delta H^\circ = -100.5\ \text{kJ} \tag{8.8}$$

and

$$CO_2 + 3\ H_2 \rightarrow CH_3OH + H_2O \qquad \Delta H^\circ = -61.6\ \text{kJ} \tag{8.9}$$

Both these reactions occur reversibly but, as discussed in Chapter 7, acceptable yields can be obtained using staged reactors with interbed cooling or similar concepts. There has been much discussion in the literature over the nature of the active catalyst and its surface under reaction conditions and also concerning the importance of the catalyst composition. It is generally accepted that the CO_2 is necessary for good activities (the conversion of pure CO is lower than that of mixtures of CO and CO_2) and it has been suggested that the CO_2 causes the surface of the catalyst to be partially oxidized under operation conditions. More recent work has shown that the conclusion that surface oxide species are present under reaction conditions was based on misleading evidence and that it is most likely that the active surface is metallic copper. Nevertheless, the CO_2 addition is necessary and it is possible that this is because its presence is related to the participation of formate species in the reaction mechanism, these species being more easily formed from the CO_2 than from CO.

TASK 8.3 The Active Site in Methanol Synthesis Catalysts

Use the paper by Klier noted above (Adv. Catal. (31 (1982) 243–313) as a source for a citation search to find recent papers discussing the effect of CO_2 on methanol synthesis and also to learn more on the most up-to-date views on methanol synthesis catalysts. You should also examine the paper by G.C. Chinchen, K.C. Waugh, D.A. Whan, "The activity and state of the copper surface in methanol synthesis catalysts", Appl. Catal. 25 (1986) 101–107, and more recent citations to that paper in relation to the nature of the active sites on Cu catalysts. These citations to this will in turn lead to work by A.T. Bell and his group on Cu/ZrO_2 catalysts which address, among other things, the importance of hydrogen spillover from Cu to zirconia during methanol synthesis.

Methanol is an important chemical feedstock with an estimated worldwide demand for 2010 of 45 million metric tons. The following web link gives details of the most important uses of methanol: http://www.methanol.org/pdf/WorldMeOHDemandbyDerivative.pdf\. This comes from a very useful source of information on methanol utilisation, the Methanol Institute (http://www.methanol.org/). Two of the most important uses of methanol are in the production of formaldehyde (Figure 8.1, reaction 8; see next section) and of methyl tertiary butyl ether (MTBE – Box 8.4). A current relatively limited use is in fuel cell operation. However, there has been a very intense period of research on that subject and so we will now consider this topic briefly.[10]

TASK 8.4 Synthesis of Higher Alcohol Mixtures

Under certain circumstances and with specially formulated catalysts (e.g. promoted Cu–ZnO–alumina materials), instead of methanol, syngas can give a mixture of alcohols by what is effectively a variant of Fischer–Tropsch synthesis: a chain growth mechanism. There was an upsurge of interest in this process in the 1980s as a result of a search for new additives for petroleum which would increase the octane and also the oxygen content. A review of the literature until 1987 by X. Xiaoding, E.B.M. Doesburg and J.J.F. Scholten ("Synthesis of higher alcohols from syngas – recently patented catalysts and tentative ideas on the mechanism", Catal. Today 2 (1987) 125–170) gives an excellent lead into the subject. You should examine that publication and also the papers citing it to obtain further knowledge of the subject. (See also Box 8.4 below.)

Fuel cells have high efficiencies for the production of electricity and are being developed for use in a variety of situations.One such application is in automobile power trains in which their use would have both environmental and energy saving advantages. The most effective

[10] It is worth noting that the large-scale use of methanol reforming for the production of hydrogen in situations in which hydrogen from other sources is less easily available has been proposed by workers from Lurgi (J. Haid and U. Koss, "Lurgi's mega-methanol technology opens the door for a new era in downstream applications", in: E. Iglesia, J.J. Spivey, T.H. Fleisch (Eds.), Natural Gas Conversion VI, Studies in Surface Science and Catalysis, Elsevier, 399–404). These authors point out that the production of hydrogen from the methanol from their new process will compete with that from naphtha.

BOX 8.4 MTBE and Higher Alcohol Mixtures

It should be noted that interest in the potential use of higher alcohols as oxygen additives dwindled in the late 1980s because of the introduction of MTBE (methyl tertiary butyl ether). This molecule was used as an oxygenated additive for petroleum (aiding in the reduction of CO emissions) for some time until it was recognized that its use presented serious environmental problems and it is now being phased out once more. (http://auto.howstuffworks.com/fuel-efficiency/fuel-consumption/question347.htm\) The production of MTBE, using methanol as a reactant, is also a catalytic process carried out using an acid catalyst in a "reactive distillation" system. The literature on this subject can be easily accessed by doing a Google search on MTBE.

fuel for use in such applications would be pure hydrogen but its use is still not effective because of the difficulties in distribution and storage, the latter particularly in the automobile itself because of the need for either high pressures or a suitable hydrogen storage material. Methanol is considered to be a promising alternative which would be used as a source of pure hydrogen or even directly in the fuel cell ("direct methanol fuel cell", DMFC). The fuel cells used can be of either low-temperature or high-temperature designs (see, e.g., C. Song, "Fuel Processing for low-temperature and high-temperature fuel cells – challenges and opportunities for sustainable development in the century", Catal. Today 77 (2002) 17–49). One of the most highly developed fuel cells for such an application is the "proton-exchange membrane fuel cell" (PEMFC) illustrated schematically in Figure 8.7. The membrane through which the protons diffuse is a fluorinated sulfonic acid polymer such as Nafion (DuPont); this is very selective for the transport only of H+ ions. Hydrogen is generally produced in the fuel processor (rather

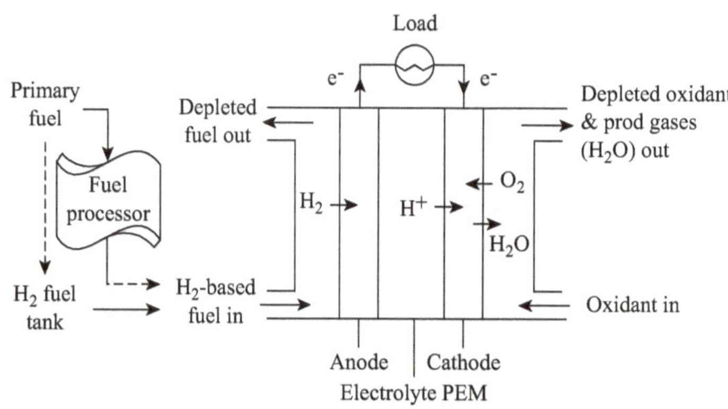

Anode (Fuel) reaction:	$H_2 = 2H^+ + 2e^-$
Cathode (Oxidant) reaction:	$1/2\ O_2 + 2H^+ + 2e^- = H_2O$
Total reaction:	$H_2 + 1/2\ O_2 = H_2O$

Figure 8.7:
Schematic representation of the arrangement of a proton-exchange membrane fuel cell (PEMFC). *Reproduced with permission from C. Song, Catal. Today 77 (2002) 17–49. Reproduced with kind permission of Elsevier.*

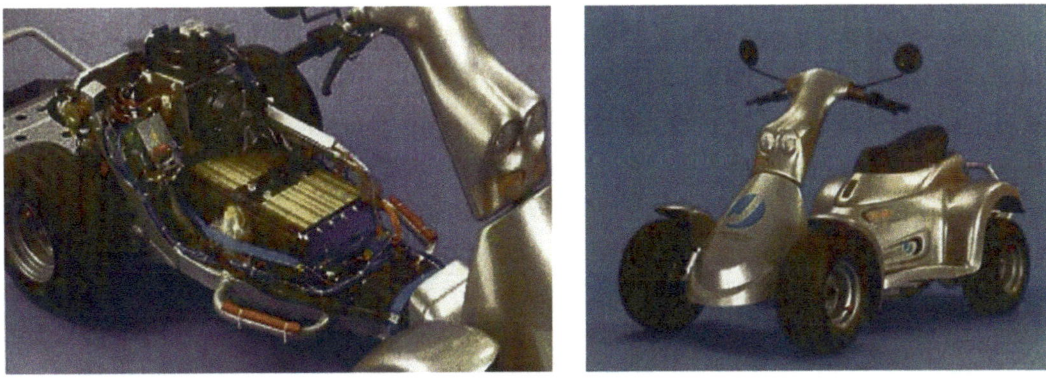

Figure 8.8:
A direct methanol fuel cell vehicle (JuMOVe) developed at the Research Centre Jülich. *Taken with permission from Forschungscentrum Juelich GmbH from P. Biedermann, T. Grube, B. Höhlein, Methanol as an energy carrier, Schriften des Forschungscentrum Jülich/Energy Technology, vol. 55, (2006). ISSN 1433–552.*

than being supplied from external sources as hydrogen gas) and passes to the anode of the fuel cell where it is converted into protons. These then pass through the membrane to the cathode where they react with oxygen from the air to produce water. Figure 8.8 shows one application of a direct methanol fuel cell (DMFC).

A wide variety of catalysts have been tested for the production of hydrogen from a number of different fuel molecules. However, the main work appears to have been done using methanol steam reforming:

$$CH_3OH + H_2O \rightarrow CO_2 + 3\ H_2 \tag{8.10}$$

The catalysts which have received most attention for this reaction (see Task 8.5) seem to have been based on formulations containing copper, these being similar to those used for the methanol synthesis reaction.

As an example of such work on the methanol reforming reaction, we will now consider some of the results obtained in a study carried out by J.P Breen and J.R.H. Ross as part of a project aimed at developing a ship powered by a fuel cell as this illustrates some of the important parameters which need to be considered in developing such catalysts ("Methanol reforming for fuel-cell applications: development of zirconia-containing Cu–Zn–Al catalysts", Catal. Today 51 (1999) 521–33). In particular, we demonstrate the importance of carrying out the relevant thermodynamic calculations when carrying out such work. Figure 8.9 shows typical data obtained for sequentially precipitated $Cu-ZnO-Al_2O_3-ZrO_2$ catalyst as a function of reaction temperature and compares them with thermodynamically calculated concentrations (see Box 8.5) for a situation in which CO formation was not allowed (lines labelled *m*) or was allowed (lines labelled *n*). It can be seen that methanol conversion begins to occur at temperatures as low as 130 °C and that it is complete at 340 °C. The maximum possible percentage of hydrogen formed

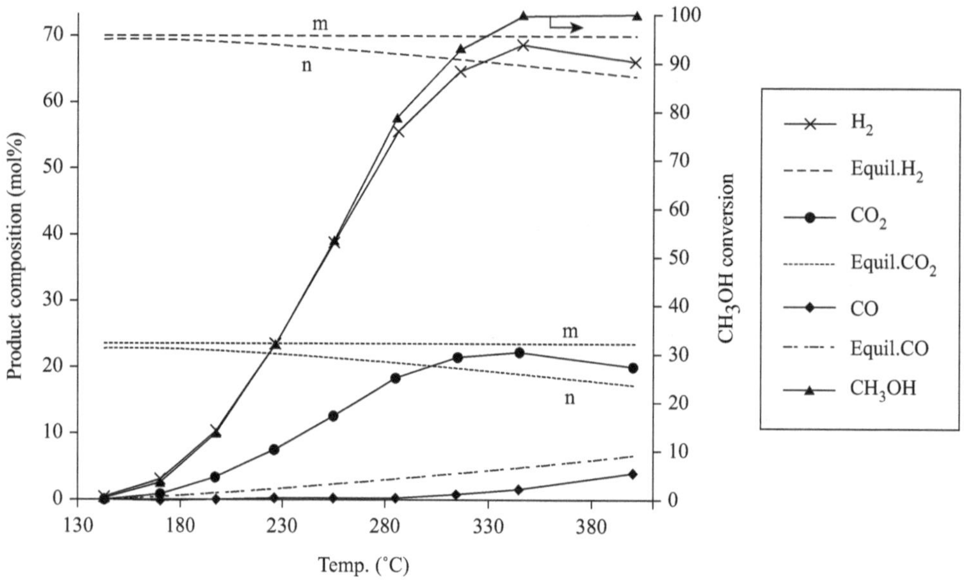

Figure 8.9:
Steam reforming of methanol over sequentially precipitated Cu catalysts. The influence of temperature on the product compositions for a $Cu-ZnO-ZrO_2-Al_2O_3$ catalyst prepared by sequential precipitation as a function of reaction temperature. The data are compared with the product mix which would have been expected thermodynamically if CO was not (lines *m*) or was (lines *n*) allowed as a product in the reaction. (Reaction conditions: $H_2O/CH_3OH = 1.3$, $P = 101$ kPa, $W/F = 0.00259$ g min^{-1} cm^{-3}). *Reproduced, with kind permission from Elsevier, from J.P. Breen, J.R.H Ross, Catal. Today 51 (1999) 521–533.*

if no CO can be formed is just below 70% and the corresponding percentage of CO_2 is about 23% (balance H_2O not shown). The experimentally determined value of the hydrogen and CO_2 contents approached the thermodynamically calculated values (see Box 8.5) at the temperature at which the methanol conversion was 100% (at about 540 °C) and then dropped off slightly at a higher temperature; the concentrations of both products were well above those permitted if CO was formed in equilibrium. It was therefore concluded that the primary reaction in the steam reforming of methanol is that shown in Eq. (8.10) and that the reverse water–gas shift reaction:

$$CO_2 + H_2 \rightleftarrows CO + H_2O \tag{8.11}$$

occurs only once the conversion of methanol approaches 100%.[11]

[11] It is interesting to note that contrasting conclusions have been reached in a more recent paper on the use of very similar catalysts, but prepared by coprecipitation rather than by sequential precipitation, for the oxidative steam reforming of methanol; that is, the same reaction but with the addition of oxygen to the feed. S. Velu, K. Suzuki, M. Okazaki, M. P. Kapoor, T. Osaki, F. Ohashi, "Oxidative steam reforming of methanol over CuZnAl(Zr)-oxide catalysts for the selective production of hydrogen for fuel cells: Catalyst characterization and performance evaluation", J. Catal. 194 (2000) 377–384.

BOX 8.5 Thermodynamic Calculations of Gas Composition

It is often desirable to know whether or not a possible reaction in a catalytic reaction is or is not in equilibrium. In the present case, it is desirable to know whether the reverse water–gas shift reaction (Eq. 8.11) is at equilibrium or not during the steam reforming of methanol. A similar approach can be used to determine whether or not the same reaction is in equilibrium during the CO_2 reforming of methane or whether or not carbon deposition is likely to occur on a catalyst. Let us consider the case shown in Figure 8.8. Two approaches are possible. One is to calculate the apparent equilibrium constant for the reaction in question using the measured partial pressures of the reactants and products for each set of experimental conditions and to compare this value with the thermodynamically calculated equilibrium constant. However, the alternative approach adopted here is to calculate for each conversion of methanol the equilibrium compositions that would be expected if the water–gas shift equilibrium was achieved and to compare these compositions with what was obtained experimentally. The calculations for Figure 8.9 were carried out using a simple thermodynamic computer program HSC Chemistry (www.outotec.com) which works out the equilibrium composition for any set of input pressures and reaction temperature. In this case, the input concentrations were calculated from the conversion of methanol and the partial pressures of water, hydrogen and CO_2 that would have resulted if no CO was formed. Similar calculations have been carried out to show the stable bulk phase of a catalyst that might be expected at any stage of a reaction, for example whether or not a carbide might remain stable in the methane + CO_2 reaction.

TASK 8.5 Methanol Reforming Catalysts

Catalysts used for the steam reforming of methanol for fuel cell use have been reviewed by C. Song ("Fuel processing for low-temperature and high-temperature fuel cells – Challenges and opportunities for sustainable development in the 21st century", Catal. Today 77 (2002) 17–49) and by A.F. Ghenciu ("Review of fuel processing catalysts for hydrogen production in PEM fuel cell systems", Curr. Opin. Solid State Mater. Sci. 6 (2002) 389–399). Use these reviews as a starting point to examine the literature on methanol reforming catalysts, paying particular attention to any significant advances in catalyst formulations and to current views on reaction mechanism.

8.2.4 Methanol Conversion to Other Products

The oxidation of methanol to formaldehyde (Figure 8.1, reaction 8) was discussed in Section 3.4.2 and hence it will not be discussed further here. In passing, it is worth mentioning that the direct conversion of methane to formaldehyde has also received a great deal of attention. However, although it is claimed that the space-time yields are high for this process, the conversions and selectivities are very low.[12] If it was possible to carry out this or the direct conversion of methane to methanol with high yields (both reactions being thermodynamically favourable but hindered by parallel total oxidation to CO and CO_2), these conversions would clearly have very significant economic advantage over the indirect routes currently used.

The conversion of methanol to higher hydrocarbons (Figure 8.1, reaction 9) over Mobil's ZSM-5 zeolite was the basis of the so-called methanol to gasoline (MTG) process.[13] This process was developed following the energy crisis caused by the Arab oil embargo and the only plant was built in New Zealand, operating only for a relatively short time after its start-up in 1985. A full description of the plant and its operation are given in a paper entitled "MTG Revisited" by C.D. Chang (Natural gas conversion, A. Holmen, K.-J. Jens, S. Kolboe (Eds.), Stud. Surf. Sci. Catal. 61 (1991) 393–404) who also gives his views on the mechanism of the process. In essence, the reaction occurs in two steps: the conversion of a mixture of methanol and dimethyl ether to olefins (with a predominance of propylene) and the subsequent conversion of the olefins to higher hydrocarbons, including aromatics. Essential to the process is the acidity of the zeolite (brought about by the presence of acidic Si-OH^+–Al groupings on the surface) which allows the formation of surface carbene (CH_2=) species which can in turn insert into C–O bonds:

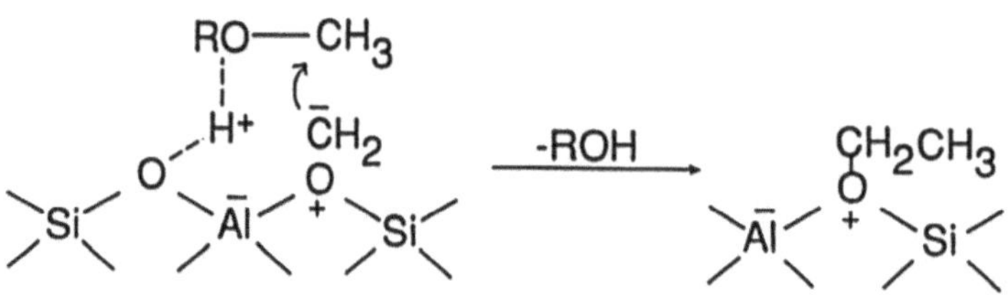

(8.12)

or undergo methylation:

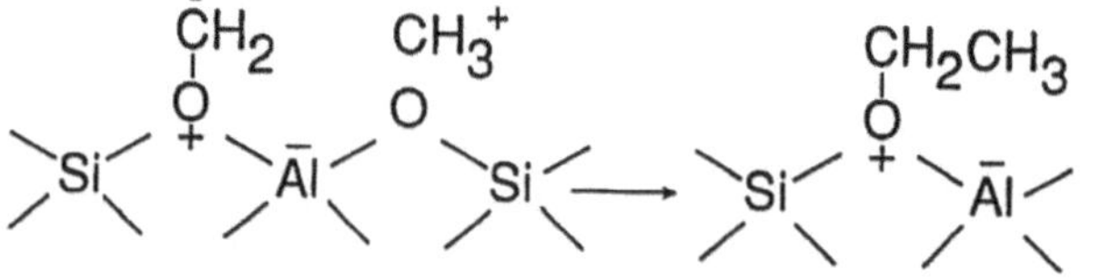

(8.13)

Once the first C–C bond is formed, the other products are formed by further insertion reactions.[14] The economics of the process were such that there was little incentive to construct

12 The conversion of methane to formaldehyde is discussed, among other topics, in a valuable and highly cited review written by J.H. Lunsford, "Catalytic conversion of methane to more useful chemicals and fuels: a challenge for the 21st century", Catal. Today 63 (2000) 165–174.

13 The first publication on the subject dates from 1977: C.D. Chang, A.J. Silvestri, J. Catal. 47 (1977) 249.

14 For further details on the mechanism, the reader should refer to the paper by Chang referred to above and also some of the papers cited in that reference. Essential to the mechanism is also the open pore structure of ZSM-5.

similar plants elsewhere; indeed, it is reported that the New Zealand plant is now used only to manufacture methanol. It is interesting to note, however, that a variant of the process, to produce predominantly propylene (methanol to propylene, MTP), is now offered by Lurgi.[15] In the Lurgi process, crude methanol is partially converted to dimethyl ether in a DME pre-reactor and the mixture is then converted over a "selective tailor made zeolite" in the presence of steam at 400–450 °C. Due to the increasingly high cost of propylene, an important chemical intermediate, the process economics is considered to be favourable if cheap methanol is available. We will return briefly to a discussion of the other routes to olefins shown in Figure 8.1 in the following section.

8.3 Catalysis in the Conversion of Crude Oil

In the early part of the twentieth century, the chemical industry relied largely on coal as the main source of chemical products and it was only with the increasing supplies of crude oil available in America and later in the Middle East and Europe that the modern petroleum and petrochemical industries came into their own. The oil industry is now the world's largest business, supplying a vast proportion of the world's energy demands as well as the majority of the petrochemical intermediates needed for the production of a wide variety of the goods that we all now take for granted: polymers, pharmaceuticals, dyestuffs, etc. Catalysis plays an important role in almost all processes carried out in the oil industry. Many of these processes are now well established and so the amount of research carried out on them relative to other areas of interest has decreased significantly in the last 20–30 years. Nevertheless, there is still a great need to improve the efficiencies of the processes used as even a small percent increase in the efficiency of a process can give enormous economic benefit by giving a relative decrease in the consumption of crude oil.

This section does not attempt to cover the whole of this very vast subject. Instead, the aim is to give a flavour of some of the most important reactions and to give pointers for further reading.

8.3.1 Crude Oil Distillation and Feedstock Cleaning

Figure 8.10 gives a very schematic representation of the arrangement of a typical oil refinery. The feed for the process, a crude oil, enters at the left-hand side where it is separated by distillation and vacuum distillation into a variety of different fractions ranging from gases to asphalt. Since the crude oil, depending on its origin, may contain a wide range of different components and impurities, each refinery will be different, dependent on that source. In the refinery depicted in Figure 8.10, several of the fraction streams are passed through *Merox units* to remove mercaptans. The Merox process, introduced by UOP (http://en.wikipedia.org/wiki/Merox/), involves reaction of the mercaptans with alkali to give organic disulfides

[15] See reference in footnote 10.

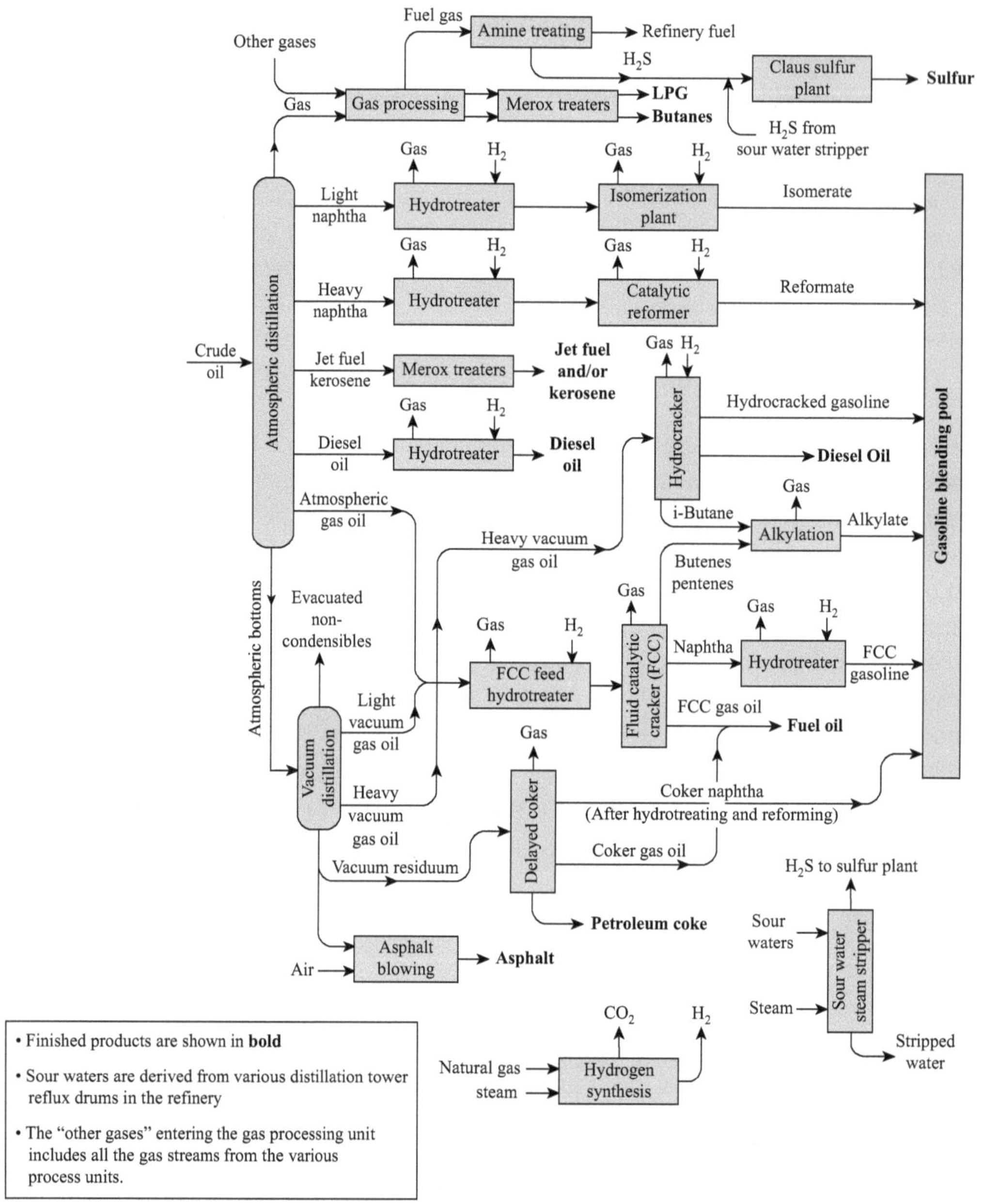

Figure 8.10:
Schematic representation of the processes in a typical petroleum refinery. *Source: http://en.wikipedia.org/wiki/File:RefineryFlow.png*

BOX 8.6 The Claus Process

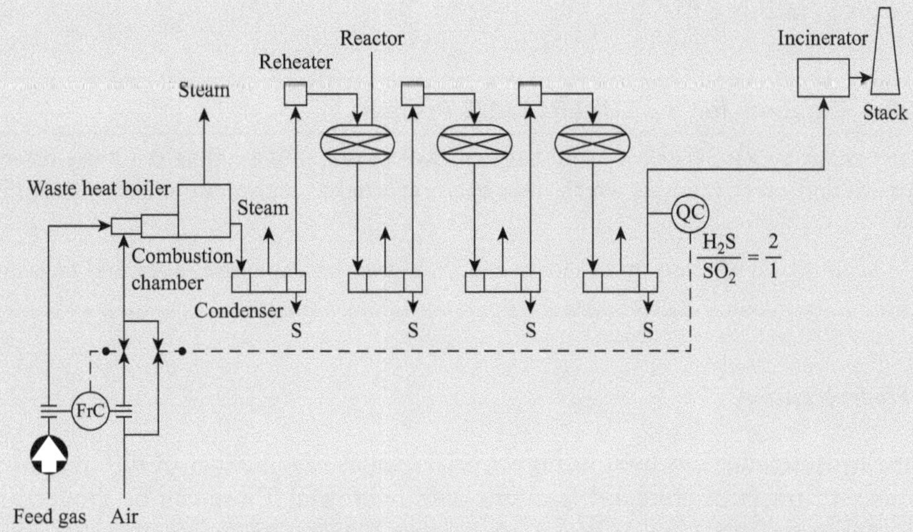

The Claus process, depicted schematically above, is essentially the reaction:

$$2\ H_2S + SO_2 \rightleftarrows 3S + 2\ H_2O$$

This can be catalysed by alumina or titania.

The SO_2 required for the reaction is prepared in a pre-burner by the reaction:

$$2\ H_2S + 3\ O_2 \rightarrow 2\ SO_2 + 2\ H_2O$$

The all-over reaction is therefore:

$$2\ H_2S + O_2 \rightarrow S_2 + 2H_2O$$

The first reaction is equilibrium limited and gives a conversion efficiency of only 95–98%, leaving significant concentrations of H_2S in the gas stream. Many Claus plants have therefore been retrofitted by a Super-Claus reactor containing a catalyst comprising an α-alumina support impregnated with iron and chromium oxides. This catalyst selectively oxidizes the majority of the remaining H_2S to sulfur at 200–300 °C without the formation of any further SO_2, giving an efficiency of 99%.

Reference: P.F.M.T. van Nisselrooy, J.A. Lagas, SUPERCLAUS reduces SO_2 emission by the use of a new selective oxidation catalyst, Catal. Today 16 (1993) 263–271.

Source: Diagram reproduced from Catal. Today 16 (1993) 265 with kind permission of Elsevier.

that are then removed from the system. The sulfur in several other streams is removed by hydrotreating (see Section 8.3.2 below). Hydrogen sulfide, which may be present in the gaseous fraction of the crude oil and also formed as a product of hydrotreating, is removed by amine scrubbing followed by a Claus plant (http://en.wikipedia.org/wiki/Claus_process\).

As discussed in Box 8.6, more stringent requirements have led to the introduction of the SUPERCLAUS process in which residual H_2S is selectively oxidized to sulfur (see the reference given in Box 8.6).

TASK 8.6 Catalysts for the SUPERCLAUS Process

Using the paper by Van Nisselrooy and Lagas shown in Box 8.6, examine the more recent literature to find other catalysts which have been reported as active in the SUPERCLAUS process.

Note. In Scopus, you will find that these authors are given as Van Nisselrooya and Lagasb!

8.3.2 Hydrotreating

The term "hydrotreating" as used in Figure 8.10 applies to a number of different processes that involve treatment of crude oil fractions with hydrogen. These can be grouped in two sub-classifications.[16] The first of these, often termed "hydropurification", covers the removal from the feedstock molecules of sulfur, nitrogen, oxygen and/or heavy metals (e.g. Ni, V, As, etc.), the corresponding treatment processes being known as hydrodesulfurization (HDS), hydrodenitrogenation (HDN), hydrodeoxygenation (HDO) and hydrodemetallization (HDM). The removal of these components means that the product molecules have improved properties and do not emit contaminants on combustion; the products can also be further treated catalytically without poisoning the catalysts used. The second group of treatments, often referred to as "hydroconversion", involves changes in the molecular weights and structures of the treated molecules in processes such as hydrogenation (HYD), hydrodearomatization (HDA), isomerization (ISM) and hydrocracking (HCG). We will first deal briefly with the process of hydrodesulfurization and will then discuss a number of different hydroconversion processes.

TASK 8.7 Active sites for the HDS Process

Grange and Vanhaeren[16] list some 37 reviews on the subject of hydrotreatment published from 1957 onwards. Using some or all of these as search terms in Scopus or Web of Science, identify and list in tabular form some more recent reviews, giving a brief description of the content of each. Using the Grange and Vanhaeren review and some of the more recent reviews, discuss the various models postulated for the active sites involved in catalysts for the HDS process.

[16] See, for example, P. Grange, X. Vanhaeren, "Hydrotreating catalysts, an old story with new challenges", Catal. Today 36 (1997) 375–391.

8.3.2.1 Hydropurification

Of the various hydropurification processes, the one which has received most attention in the literature is hydrodesulfurization. The catalysts for hydrodesulfurization, generally based on the Co/Mo/alumina system, are often made using preformed γ-Al_2O_3 extrudates by a simple impregnation method involving, for example, solutions of cobalt nitrate or formate and ammonium molybdate. One of the simplest S-containing molecules used for model studies is thiophene. Hargreaves and Ross examined the relative rates of the HDS of thiophene and tetrahydrothiophene as well as of the hydrogenation of butadiene and but-1-ene over a catalyst with typical Co and Mo contents which had been pre-sulfided[17] and that the results could be explained by the scheme shown in Figure 8.11. The authors argued from their results that the parallel scheme in which the first product is but-1-thiol (i.e. ring opening before desulfurization) did not occur to any appreciable extent; they suggested that the predominant route is the hydrogenation of the double bonds in adsorbed thiophene species prior to the cleavage of C—S bonds in the adsorbed tetrahydrothiophene to give but-1-ene as the predominant preliminary product. These conclusions contrasted with those reached earlier by Amberg and his colleagues who thought that the formation of but-1-thiol was the first step in the process.[18] However, the difference between the two sets of conclusions may largely be due to the different experimental conditions used in the two studies. Although this work was carried out more than 30 years ago, there is still discussion on sequence of surface reactions. For example, Henrik Topsøe and his colleagues at Haldor Topsøe, who have added a great deal to the literature on

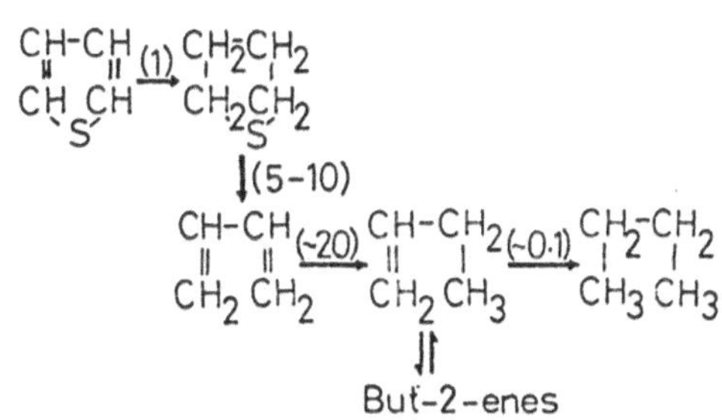

Figure 8.11:
Scheme for thiophene hydrogenation to butenes and butane; the relative rates are given above the arrows for a catalyst containing 4.9 wt% CoO and 11.6 wt% MoO_3 with a balance of γ-Al_2O_3. The catalyst was pre-sulfided in H_2S at 573 K for 17 h and then in H_2 at 521 K for 1 h. *Scheme taken from Hargreaves and Ross: reference in footnote 17. Reproduced with kind permission of Elsevier.*

[17] A.E. Hargreaves, J.R.H. Ross, "An investigation of the mechanism of the hydrodesulfurization of thiophene over sulfided Co-Mo/Al_2O_3 catalysts. II. The effect of promotion by cobalt on the C—S bond cleavage and double bond hydrogenation/dehydrogenation activities of tetrahydothiophene and related compounds", J. Catal. 56 (1979) 363—376.

[18] See, for example, P. Dekisan, C.H. Amberg, "Catalytic hydrodesulfurization of thiophene. 5. Hydrothiophenes. Selective poisoning and acidity of catalyst surface", Canad. J. Chem. 42 (1964) 843.

the HDS reaction, have carried out a series of computations on the energetics of various possible surface steps on an idealized MoS_2 surface: see Task 8.8. The hydrodenitrification (HDN) and hydrodemetalization (HDM) processes are very similar to the HDS process, although the catalysts used may be slightly different. For example, HDN often uses a Ni–Mo–alumina catalyst and HDM requires catalysts with wide pores which are not poisoned too seriously by the deposition in them of the metallic impurities.

TASK 8.8 Computational Approach to HDS Mechanisms

Moses et al. have carried out a series of computations on the steps which occur during the hydrodesulfurization of thiophene over the active sites on MoS_2 and have shown the relevant importance of the various possible steps on S and Mo edge sites as shown in the scheme below:

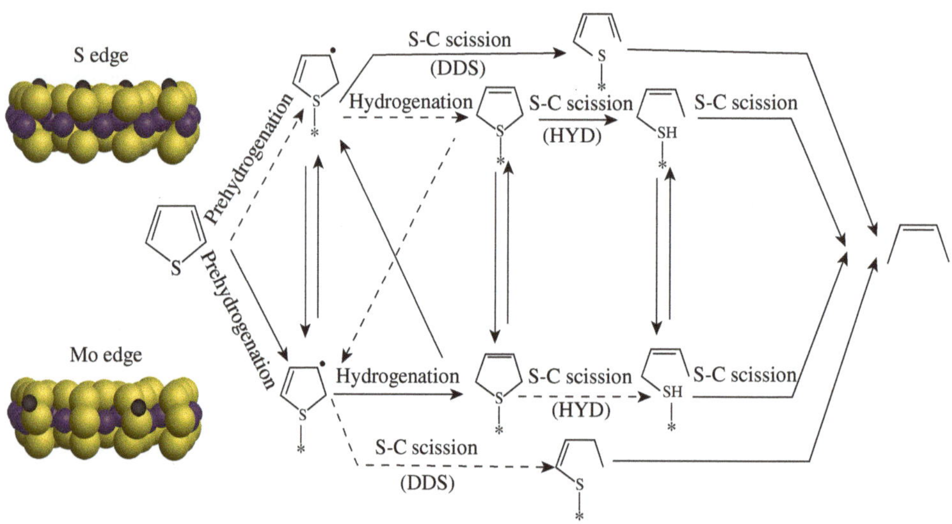

They have shown in the corrigendum to the original paper that the S-C scission (DDS – direct desulfurization) route on S-edge sites at the top of the diagram is more important than the HYD (hydrogenation) route on Mo sites at the bottom. The predominant route is therefore likely to be prehydrogenation on the Mo edge sites followed by transfer to the S-edge sites and S-C scission on those sites. Use the references in the main paper to gain an understanding of the background to this work and the evidence for the site geometry discussed.

- P.G. Moses, B. Hinnemann, H. Topsøe, J.K. Nørskov, "The hydrogenation and direct desulfurization reaction pathway in thiophene hydrodesufurization over MoS_2 catalysts at realistic conditions: A density functional study", J. Catal. 248 (2007) 188–203.
- P.G. Moses, B. Hinnemann, H. Topsøe, J.K. Nørskov, Corrigendum to the previously cited paper, J. Catal. 260 (2008) 202–203.

(NOTE: It is not suggested that you attempt to understand all the details of the computational methods used, although some readers may find this very instructive.)

Diagram reproduced with kind permission of Elsevier.

TASK 8.9 HDS, HDN and HDM Catalysts and Processes

Using as a source reference the following very comprehensive review by E. Furimsky and F.E. Massoth, examine the literature on HDS, HDN and HDM catalysts and processes: "Deactivation of hydroprocessing catalysts", Catal. Today 52 (1999) 381–495.

You should also pay attention to a more recent paper by the same authors on the HDN process: "Hydrodenitrogenation of petroleum", Catal. Rev. 47 (2005) 297–489.

8.3.3 Isomerization of Light Naphtha

Light naphtha (see Figure 8.9) contains a small proportion of butane and larger fractions of *n*-pentane and *n*-hexane. As they stand, the latter molecules have relatively little value as fuel additives because of low RON numbers.[19] However, if they are isomerized to give 2-methyl butane and 2,2-dimethyl butane respectively, RON values of 99 and 89 are achieved. The aim of the isomerization step shown is therefore to bring about such an isomerization reaction. While the isomerization of alkenes is relatively easily achieved, that of the alkanes is not. Hence, it is necessary for the catalysts used to have two functions: dehydrogenation/hydrogenation and isomerization. The former function is normally provided using a noble metal component, notably Pt, while the latter function is normally given by using a solid acid material. Hence the term "bifunctional catalyst". Early bifunctional catalysts consisted of a strong acid such as $AlCl_3$ promoted by Pt but these have been phased out for environmental reasons; more recently, other acid components such as alumina, silica–alumina, zeolites and sulfated zirconia have all been used, again promoted by Pt. The operation of a bifunctional catalyst for the isomerization of *n*-pentane can be depicted as follows[20]:

$$nC_5H_{12} \rightarrow nC_5H_{10} + H_2 \text{ (on Pt)} \tag{8.15}$$

$$nC_5H_{10} + H^+ \rightarrow nC_5H_{11}{}^+ \text{ (on solid acid)} \tag{8.16}$$

$$nC_5H_{11}{}^+ \rightarrow \text{iso}C_5H_{11}{}^+ \text{ (on solid acid)} \tag{8.17}$$

$$isoC_5H_{11}{}^+ \rightarrow isoC_5H_{10} + H^+ \text{ (on solid acid)} \tag{8.18}$$

$$isoC_5H_{10} + H_2 \rightarrow isoC_5H_{12} \text{ (on Pt)} \tag{8.19}$$

19 The research octane number (RON) gives the relative ability of a molecule to resist reaction in an internal combustion engine in an oxygen-containing atmosphere until the conditions are at those required for ignition. Molecules with low values will give "knocking" or "pinking" while those with high values will not. The molecule iso-octane (2,2,4-trimethylpentane) is taken to have a value of 100. In the past, a molecule such as tetra-ethyl lead was added to the mixture to prevent knocking but this was phased out several decades ago.

20 This scheme is taken from an excellent review on the subject by Y. Ono, "A survey of the mechanism in catalytic isomerization of alkanes", Catal. Today 81(2003) 3–16.

The isomerization reaction is generally carried out in the presence of hydrogen as this helps to retain catalyst activity over longer periods without the need for frequent regeneration as a result of coke formation. When required, carbon burn-off is achieved using a mixture of steam and air.

TASK 8.10 Use of Mo Carbide Catalysts in Hydrocarbon Isomerisation

Using the reference given in footnote 20 as a starting point, examine the potential use of Mo carbide catalysts for hydrocarbon isomerization.

8.3.4 Reforming of Heavy Naphtha

Heavy naphtha, a fraction of crude oil that boils between approximately 70 and 200 °C, contains mostly alkanes and cycloalkanes although there may also be a small fraction of aromatics. This mixture has a relatively low RON value and so the reforming process is aimed at increasing that value by a mixture of different reactions such as dehydrogenation, isomerization, dehydrocyclization (ring closure followed by increasing the aromatic content by dehydrogenation) and some hydrocracking to reduce the chain length of the heavier molecules present. As with the isomerization process discussed above, the catalysts used must have a dehydrogenation function to produce olefins and an acid function to bring about isomerization; the metallic function also first gives some limited cracking of the heavier molecules which are then isomerized. Much has been written about the catalysts, reaction schemes and processes for naphtha reforming and the research literature on the subject is vast. According to Bond in his seminal book on "Metal-catalysed reactions of hydrocarbons" (Springer 2005), which the reader should peruse to gain a very full understanding of the subject, bimetallic catalysts such as Pt–Sn, Pt–Ir and Pt–Re have had widespread use in hydrocarbon reforming. It is generally accepted that one of the main functions of the added metal promoter is to stabilize small Pt crystallites and thereby to decrease the chances of carbon deposition. As with light naphtha isomerization, the catalysts still need to be regenerated regularly. Performing Task 8.11 will give the reader a feel for the subject.

TASK 8.11 Bimetallic Reforming Catalysts

Two important articles on this subject from the period around 1980 are the following:

- F.M. Dautzenberg, J.N. Helle, P. Biloen, W.M.H. Sachtler, "Conversion of n-hexane over monofunctional supported and unsupported Pt–Sn catalysts", J. Catal. 63 (1980) 119–128
- V. Ponec, "Catalysis by alloys in hydrocarbon reactions", Adv. Catal. 32 (1983) 149–214.

Use these references as source material to examine work by these authors (and by others referred to in these articles or citing them) on the mechanisms of hydrocarbon reactions on bimetallic catalysts. Look out for names such as J.H. Sinfelt, J.K.A. Clarke, R. Burch, if necessary including their names in further searches. Your work should aim to list the catalysts which have been studied and to gain an idea of the mechanisms of the reactions and of the effects of alloying on carbon deposition kinetics.

8.3.5 Fluid Catalytic Cracking (FCC)

Catalytic cracking is now one of the most important processes practiced in the petroleum refining industry as it allows the use of a much larger fraction of crude oil, converting relatively high molecular weight materials into high octane fuels. In its earliest development (http://en.wikipedia.org/wiki/Fluid_catalytic_cracking\), Friedel Crafts catalysts ($AlCl_3$) were used but the process using them was too expensive. In 1922, two Frenchmen named E.J. Houdry and E.A. Prudhomme set up a plant to produce 60 tons per day of petroleum from lignite using a catalyst discovered by Houdry, Fuller's Earth containing silica—alumina. Following further work on oil conversion, Houdry was invited by the Vacuum Oil Company in 1930 to come to the US and there he worked first with a new company set up between the Vacuum Oil Company and Standard Oil of New York (Socony), the Socony—Vacuum Oil Company, which operated the first plant, producing 32,000 litres per day. Then, working in collaboration with the Sun Oil Company, the first demonstration unit was operated in 1936 in Paulsboro, New Jersey, producing ten times the volume of that produced by the Socony—Vacuum Oil Company. By 1940, there were 14 full-scale units in operation, these being fixed-bed semi-batch systems. The FCC process as we know it arose from work in a consortium, Catalytic Research Associates (CRA), made up of Standard Oil of New Jersey, Standard Oil of Indiana, Anglo-Iranian Oil, Texas Oil, Dutch Shell, M.W. Kellogg, Universal Oil Products and IG Farben, this being based on work by Standard Oil of New Jersey prior to the Depression years. The first plant based on the new technology, built by M.W. Kellogg Company, was commissioned in 1940 at the Standard Oil of New Jersey (later Exxon Mobil) Baton Rouge refinery in Louisiana.[21] Further details of some of the history of these important developments are to be found on the North American Catalysis Societies' web site at http://www.nacatsoc.org/edu_info.asp.

Figure 8.12 is a schematic representation of a typical modern fluid catalytic cracking plant, showing the surrounding peripheral units as well as the reactor, while Figure 8.13 gives more details of the central "riser" reactor itself. (Riser reactors were also described briefly in Section 5.5.) Fresh catalyst is fed to the riser section in which the acid-catalysed cracking reaction takes place. The cracked product is then removed for further processing in the cyclone vessel. The catalyst, which now contains substantial quantities of coke, is then fed to the regenerator which is also fed with combustion air to burn-off the coke. (Other additives can be added at this stage, these including de-NO_x components designed to reduce any NO_x formed in the regenerator; see Harding et al., reference in Task 8.12.) The regenerated catalyst is then fed back to the riser where additional fresh catalyst is also added. In addition to having high activity for the cracking reaction, the catalyst must have high strength to resist degradation in the fluidized beds and to that end, many other constituents are added; see Task 8.12.

[21] High octane fuel made in the early plants was provided to the RAF for use during the Battle of Britain and it has been claimed that the superior performance achieved assisted victory in that battle.

TASK 8.12 FCC Catalysts

FCC catalysts are nowadays based on zeolitic materials together with binders and promoters of various sorts. Examples of the vast body of information available on the development of the catalysts used in the process are papers by P. O'Connor, P. Imhof, S.J Yanik, "Catalyst assembly technology in FCC, Part 1. A review of the concept, history and developments", in: M.L. Occelli, P. O'Connor (Eds.), Studies in Surface Science and Catalysis, 131 (2001) 299–310 and R.H. Harding, A.W. Peters, J.R.D. Nee, "New developments in FCC catalyst technology", Appl. Catal. A General 221 (2001) 389–396. Use these as starting points to learn more about the types of catalyst used in the process, the nature of the active sites and their function, and the use of various additives to improve environmental and other aspects of the process.

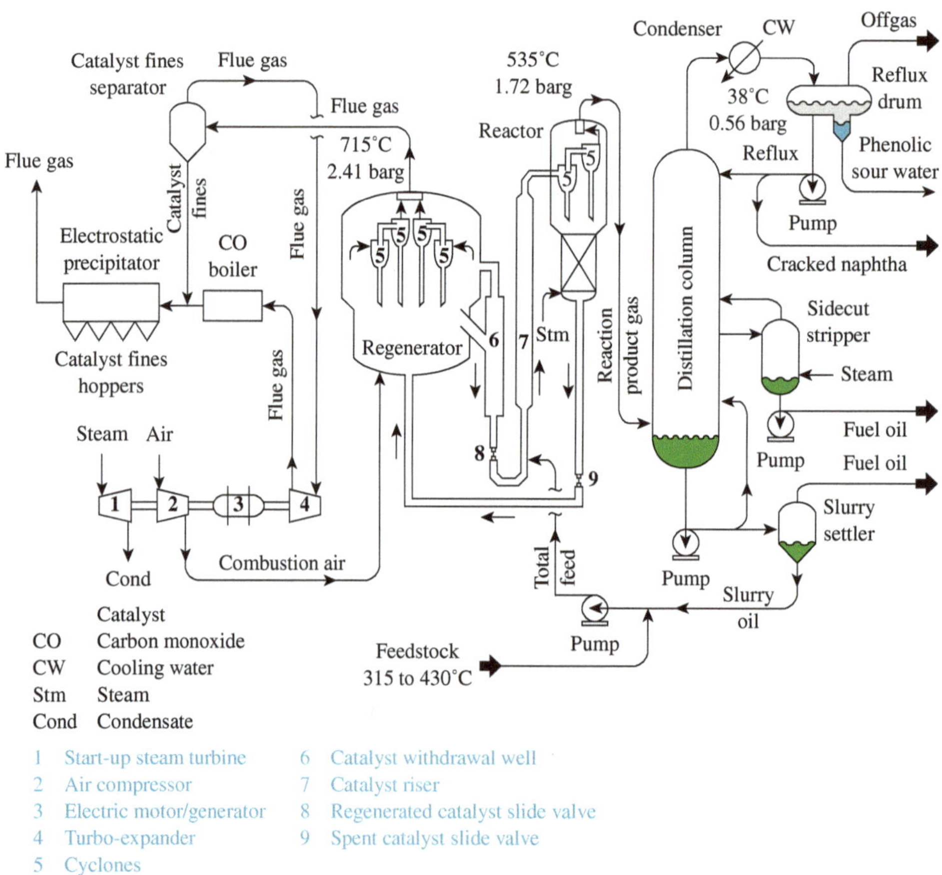

Figure 8.12:

Schematic representation of a typical FCC plant. *Reproduced from Wikipedia: http://upload.wikimedia.org/wikipedia/commons/9/95/FCC.png*

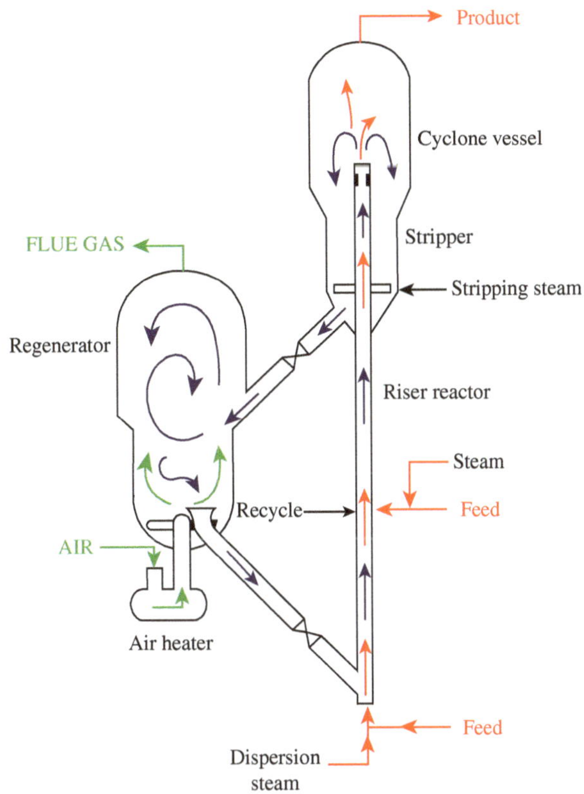

Figure 8.13:
Schematic representation of a riser reactor. The legend should include a legend showing the arrow colours with hydrocarbon, catalyst and air/flue gas respectively.

Essential for the operation of a good cracking catalyst is the presence on the surface of the catalyst of strong acidic sites that can initiate the reaction by forming carbonium ions which can then react further to form carbonium ions:

$$R\text{-}CH_2\text{-}CH_3 + H^+ \rightarrow R\text{-}\overset{H}{\underset{H\;\;H}{C^+}}\text{-}CH_3 \rightarrow R\text{-}\underset{H}{C^+}\text{-}\,CH_3 + H_2 \qquad (8.20)$$

The carbonium ions can then react further in many different ways including by hydrogen transfer:

$$R\text{-}\underset{H}{C^+}\text{-}CH_3 + R'CH_2\text{-}CH_3 \rightarrow RCH_2\text{-}CH_3 + R'CH^+\text{-}\,CH_3 \qquad (8.21)$$

C–C bond scission:

$$CH_3{-}CH^{+}{-}CH_2{-}R'' \rightarrow CH_3{-}CH{=}CH_2 + R''^{+} \qquad (8.22)$$

and isomerization:[22]

$$CH\text{-}CH^{+}\text{-}CH_2\ CH_2\text{-}CH_2\text{-}CH_2\text{-}CH_3 \rightarrow CH_3\text{-}CH\text{-}\underset{\underset{CH_3}{|}}{CH^{+}}\text{ - }CH_2\text{-}CH_2\text{-}CH_3 \qquad (8.23)$$

Once a carbonium ion is formed, a chain reaction sets in and both bond cleavage and isomerization occur (forming preferentially the more thermodynamically stable products containing secondary and tertiary carbon atoms), giving products with significant improvements in octane ratings. For further details on the FCC process, the reader is referred to some of the excellent textbooks dealing with petroleum refinery processes.

8.4 Petrochemicals and Industrial Organic Chemistry

The fields of petrochemicals and industrial organic chemistry, including pharmaceutical applications, are immense and will only be given very brief treatment in this text by giving a number of appropriate references which the student is encouraged to follow up. In the past, many organic chemicals were carried out using homogeneous reaction systems and little attention was given to factors such as environmental concerns. For example, oxidation was commonly carried out using stoichiometric reagents such as permanganate and dichromate. Hence, there is currently a great deal of interest in substituting these potentially polluting technologies by heterogeneous catalytic processes.

TASK 8.13 Butane to Maleic Anhydride

An important example of catalytic oxidation is the selective oxidation of butane to maleic anhydride:

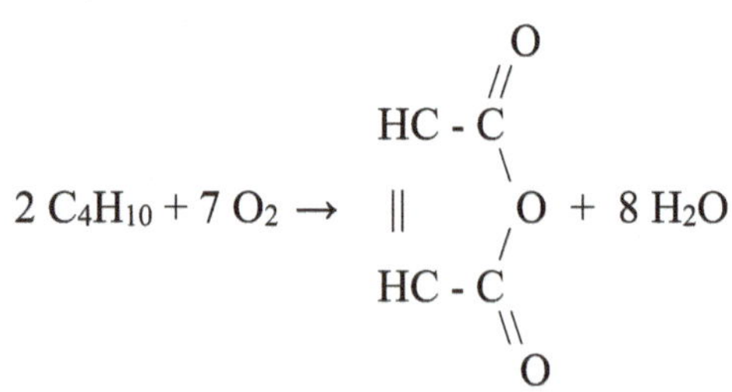

(1)

This reaction is catalysed by a specially prepared vanadium phosphate catalyst, much researched and extensively discussed in the literature. A good starting point for a literature study on this subject is a paper by G.J. Hutchings, C.J. Kiely, M.T. Sananes-Schultz, A. Burrows, J.C. Volta, "Comments on the nature of the active site of vanadium phosphate catalysts for butane oxidation", Catal. Today 40 (1998) 273–286. (See also BOX 5.6.)

[22] The isomerization process on the acidic catalyst is similar to that occurring on the acidic support in bifunctional catalysis as described in Section 8.3.3.

TASK 8.14 Oxidation Catalysis in Fine Chemicals Production

For several useful reviews of the area of oxidation, you are referred to the following references:

- R.A. Sheldon, "Heterogeneous catalytic oxidation in the manufacture of fine chemicals", in: M. Guisnet et al. (Eds.), Heterogeneous Catalysis and Fine Chemicals, II, Stud. Surf. Sci. Catal. (1991) 33–54.
- R.A. Sheldon, J. Dakka, "Heterogeneous catalytic oxidations in the manufacture of fine chemicals", Catal. Today 19 (1994) 215–245.
- R.A. Sheldon, I.W.C.E. Arends, A. Dijksman, "New developments in catalytic alcohol oxidation", Catal. Today 57 (2000) 157–166.

You should also carry out Scopus or Web of Science searches on these papers in order to identify other relevant literature.

Another set of oxidation reactions that has received a significant amount of attention over the last couple of decades is the selective oxidative-dehydrogenation of lower alkanes such as propane:

$$C_3H_8 + 1/2\ O_2 \rightarrow C_3H_6 + H_2O \tag{8.25}$$

Propylene is an important industrial intermediate, for example for the production of polypropylene, and is normally made by the non-catalytic steam cracking of lower alkanes when a mixture of ethylene and propylene is formed. As the demand for both, particularly propylene, has increased, other sources of propylene have been sought, these including the propane fraction of LPG. Reaction 8.25 is catalysed by a range of oxide materials; however, the non-selective oxidation to CO_2 and water is in most cases predominant. Although high selectivities to propylene have been obtained with a number of different catalysts, the conversions are still low and the yields are not yet competitive. The following article, which discusses the reaction over V-based catalysts, can be used as a source on which a fuller literature study can be based:

T. Blasco, J.M. López Nieto, "Oxidative-dehydrogenation of short-chain alkanes on supported vanadium oxide catalysts", Appl. Catal., A (General), 157 (1997) 117–142.

A large number of industrial organic reactions involve selective reduction reactions. These can use either homogeneous or heterogeneous catalysis. The applications of heterogeneous catalysis are many and varied, including the production of margarine and hydrogenation alkynes to alkenes. The following are the examples of source references for these two applications:

- B. Nohair, C. Especel, G. Lafaye, P. Marécot, L.C. Hoang, J. Barbier, "Palladium supported catalysts for selective hydrogenation of a sunflower oil", J. Mol. Catal., A (Chemical) 229 (2005) 117–126.

- A. Borodziński, G.C. Bond, "Selective hydrogenation of ethyne in ethene-rich streams on palladium catalysts. Part 1: Effect of changes to the catalyst during reaction". Catal. Rev. - Sci. Eng. 48 (2006) 91–144.

In fine chemical applications, it is often desirable that the reactions occur enantioselectively to produce one of two stereoisomers. The following are the examples of publications in this area:

- H.-U. Blaser, C. Malan, B. Pugin, F. Spindler, H. Steiner, M. Studer, "Selective hydrogenation for fine chemicals: recent trends and developments", Adv. Syn. Catal. 345 (2003) 103–151.
- H.-U. Blaser, H.-P. Jalett, M. Müller, M. Studer, "Enantioselective hydrogenation of α-ketones using cinchona modified Pt catalysts and related systems: A review", Catal. Today 37 (1997) 441–463.
- H.-U. Blaser, "Heterogeneous catalysis for fine chemicals production", Catal. Today 60 (2000) 161–165.

Another review from the same team deals with not only selective hydrogenation but also hydrogenolysis, dehydrogenation and oxidation: H.-U. Blaser, A. Indolese, A. Schnyder, H. Steiner, M. Studer, "Supported palladium catalysts for fine chemicals synthesis", J. Mol. Catal., A (Chemical) 173 (2001) 3–18.

Another author who has also been very active in this area of catalysis is A. Baiker (ETH Zürich) whose interests have included not only enantioselective hydrogenation but also many other topics.

8.5 Environmental Catalysis

Environmental catalysis has become very important in everyday living and to most people in the street the term "catalysis" is synonymous with car exhaust emission control. However, there are many other environmental applications, these including power station emission control, the destruction of volatile organic contaminants (VOCs) and CO, and the destruction of pollutants in water by photo-catalysis. In the early days of emission control by environmental catalysis, the appropriate papers were dotted through the literature but later specific journals appeared, notably *Applied Catalysis B* (Environmental) which was first published in1992. A number of companies have also been at the forefront in emission control, for example Engelhard (now part of BASF http://www.catalysts.basf.com/main/) and Johnson Matthey (http://www.matthey.com\). The latter company has for many years published their own journal, now available only electronically: *Platinum Metals Review*;[23] this is an

[23] The full searchable archive for *Platinum Metals Review* is available at http://www.platinummetalsreview.com/dynamic/volume/archive/. This is a very valuable source of information on all aspects of the use of the noble metals, particularly on their use in catalysts, and includes a large number of valuable review articles on the subject. As there is no link between the various articles, you will have to return to the search page to find each.

excellent source of review articles summarizing the state of the art at any stage over the last four decades and although it is unashamedly a "house journal", it is still an excellent source of background material. For example, an article by H. Windawi and M. Wyatt on "Catalytic destruction of halogenated volatile organic compounds: Mechanism of platinum catalyst systems" (Plat. Metals Rev. 37 (1993) 186–193) gives an excellent background introduction to that area of activity as well as giving a listing of earlier contributions to the same periodical on other aspects of VOC control.

8.5.1 Selective Reduction of NO_x[24]

The catalytic control of NO_x emissions from nitric acid plants and power stations, etc., using selective catalytic reduction (SCR) with ammonia as reductant has been a subject of interest for many decades and a number of commercial processes are available; see Box 8.7. The catalysts are generally composed of V_2O_5 supported on TiO_2 and washcoated onto a ceramic honeycomb matrix manufactured in the form of blocks. These blocks are then mounted in a larger structure of the type depicted in Figure 8.14. Ammonia is dosed to the system in stoichiometric quantities, depending on the NO_x content of the gas being treated. The conversions obtained approach 100% with 100% selectivity to nitrogen. The main problem with such plants is ammonia "slip" (emission of unreacted ammonia) and so the control of the ammonia addition is the most critical factor.

BOX 8.7 SCR Catalyst Manufacturers

An example of a web site giving information on a commercial catalytic system using selective reduction with ammonia from CRI Catalyst Company (part of the Shell Group) is given at http://www.cricatalyst.com/home/content/cri_catalyst/catalysts/scr_catalyst/; other companies manufacturing ammonia de-NO_x systems have the following web sites: www.basf.com/; www.coalogix.com/; www.ducon.com/; www.etecinc.net/; www.topsoe.com\; www.peabodyengineering.com/; www.seiler.co.at/; www.tenviro.com/; and www.zeeco.com/.

The mechanism of the SCR reaction over vanadia on titania catalysts using ammonia as a reductant has been a matter of some discussion; see also Box 3.6. Figure 8.15 shows schematically a possible mechanistic interpretation of the data obtained until now. There are two types of sites: one on which an oxidation–reduction process occurs (V5+ to V3+ and back) and another involving the acididic function of V5+ sites.

---

[24] NO_x is formally a mixture of the oxides of NO but is in most cases predominantly NO.

BOX 8.8 Literature on SCR Reactions, Systems, etc.

One of the earliest reviews on the subject of NO/NH3 SCR was written by H. Bosch and F.J.J.G. Janssen, "Catalytic reduction of nitric oxides: A review", Catal. Today 2 (1988) No.2.*

You can read other reviews of the subject by following some of the citations to this review or use as starting point the review by P. Forzatti, "Present status and perspectives in de-NO_x SCR catalysts", Appl. Catal. A (General), 222 (2001) 221–236.

An example of a highly cited paper on the subject of SCR catalysts is by M.D. Amarides, I.E. Wachs, G. Deo, J.-M. Jehng, D.S. Kim, "Reactivity of V_2O_5 catalysts for the selective catalytic reduction of NO by NH3: influence of vanadia loading, H_2O and SO_2", J. Catal., 161 (1996), 247–253; other papers by the same group have also been highly cited. They have also examined the use of such catalysts for the oxidation of chlorinated hydrocarbons:

- S. Krishnamoorty, J.P. Baker, M.D. Amiridis, "Catalytic oxidation of 1,2-dichlorobenzene over V2O5/TiO2 catalysts", Catal. Today 40 (1998) 39–46.

A very complete review of NO_x removal is given in the following article: V.I. Pârvulescu, P. Grange, B. Delmon, "Catalytic removal of NO", Catal. Today 46 (1998) 628.

* Note: you will probably have to look at each chapter separately as the complete issue does not download in Science Direct.

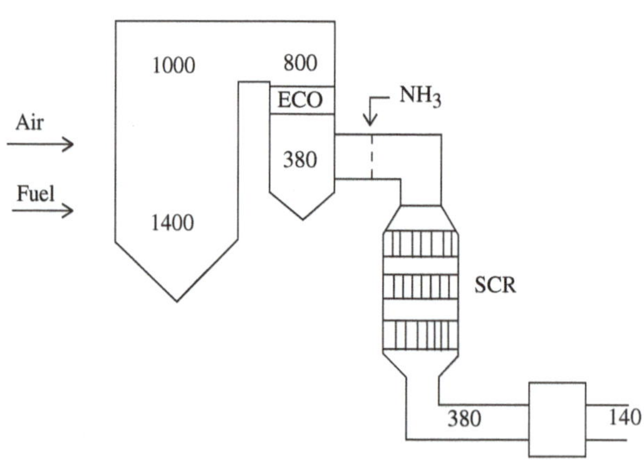

Figure 8.14:
A schematic representation of an SCR plant, showing typical temperatures (in °C) at various positions in the plant.

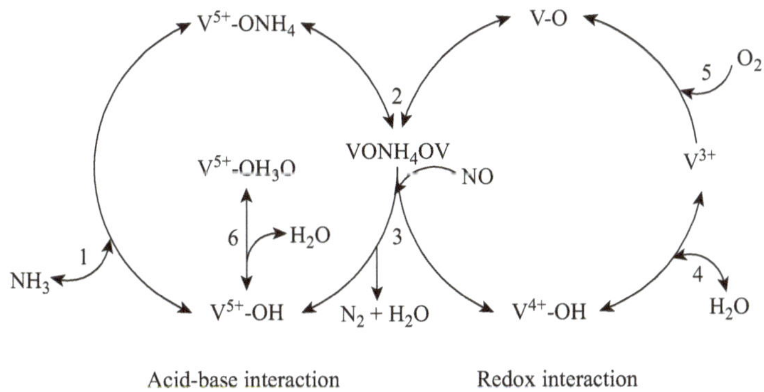

Figure 8.15:
A mechanistic scheme for the SCR reaction. *Source: Pârvulescu et al., Catal. Today 46 (1998) 233–316. Reproduced with kind permission of Elsevier.*

TASK 8.15 Mechanism of SCR

Using some of the references in Box 8.8 above, examine the mechanisms of SCR discussed and determine which, if any, is best justified by the results.

8.5.2 Automobile Exhaust Catalysts

The use of catalysts in car exhaust purification ("the catalytic muffler"), one of the most important applications of catalysis in the environmental area, is probably one of the best-known applications of catalysis of which almost everyone will have heard. It is interesting to follow the gradual developments in the requirements for three-way catalysts (i.e. control of NO_x as well as of CO and unburnt hydrocarbons) over the years, from the earliest stages following the introduction of legislation in the US demanding emission control (1970 Clean Air Amendments Act) to the present day, by reading some of the articles *in Platinum Metals Review* referred to above (see footnote 23). Box 8.9 lists some of these. Studying these articles will also give the reader a good idea of the technology applied to the control of exhaust emissions and the effect of the stoichiometry of the combustion process on the effectiveness of the catalysts used.

Figure 8.16 shows two typical forms of catalyst used in car exhaust systems: (1) ceramic honeycomb and (2) metallic monolith structures. The ceramic monolith is formed by extrusion methods using ceramic technology[25] while the metal monoliths are made (often from

[25] Many of the substrates used are made by Corning (http://www.corning.com/environmentaltechnologies/index.aspx) who also supply ceramic blocks of the type used in power stations (see previous section) and diesel filters (see below).

BOX 8.9 Articles on Car Exhaust Treatment from Platinum Metals Review

The following are some examples of the many relevant articles in PMR on exhaust catalysis:

- B. Harrison, B.J. Cooper, A.J.J. Wilkins, Development of Rh/Pt three-way catalyst systems, Plat. Metals Rev. 25 (1981) 14–22.
- B.J. Cooper, S.A. Roth, Flow-through catalyst for diesel engine emission control: Pt-coated monoliths reduce particulates, Plat. Metals Rev. 35 (1991) 178–187.
- B.J. Cooper, Challenges in emission control catalysts for the next decade, Plat. Metals Rev., 38 (1994) 2–10.
- M.V. Twigg, Twenty-five years of autocatalysis, Plat. Metals Rev. 43 (1999) 168–171.

An article by G.J.K. Acres also provides useful background knowledge: The development of emission control technology for motor vehicles, Stud. Environ. Sci., 44 (1991) 376–396.

Fecralloy, an alloy of Fe, Cr, Al and Y, originally patented by the UK Atomic Energy Authority) by a crimping process. In both cases, the matrix is then coated with a "washcoat", a slurry containing the support and the active components (see below), followed by drying and calcination. Figure 8.17 shows expanded images of two wash-coated ceramic substrates. As can be seen, the washcoating technique tends to fill the corners of the structures so that the channel is almost cylindrical. The completed monolith or honeycomb structure is then incorporated in suitable container that is generally placed between the engine and the conventional silencer module (see Figure 8.18). The whole system is controlled very carefully using sensors before and after the catalyst module in order that the fuel mix fed to the engine gives optimum behaviour in the catalyst.

We will now consider very briefly the chemistry of the reactions involved and the nature of the catalysts used. Again, the reader should consult the articles cited for more details. Critical to the operation of a catalytic exhaust control system is the air–fuel ratio of the

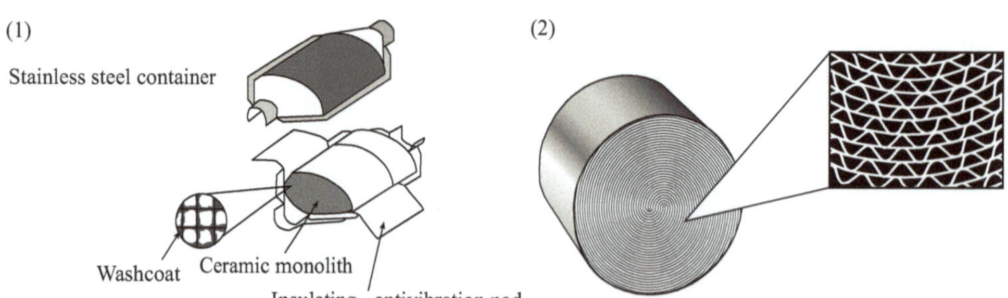

Figure 8.16:
(1) Typical catalytic converter; (2) Metal honeycomb structure (Emitech GmbH). *Source: J. Kasper et al., Catal. Today 77 (2003) 419–449. Reproduced with kind permission of Elsevier.*

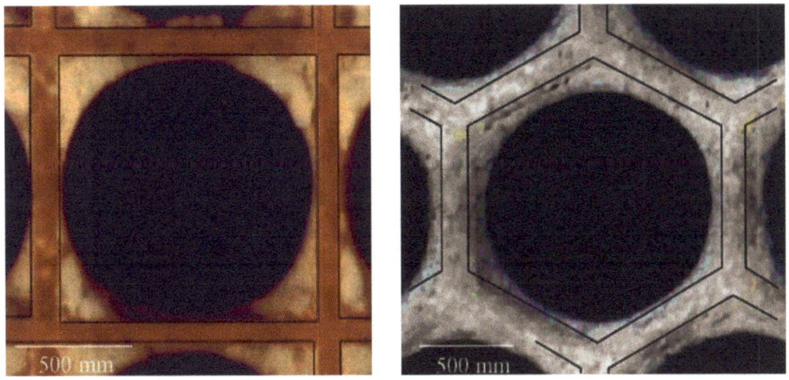

Figure 8.17:
Typical wash-coated monolith structures. *Source: S. Matsumoto, Catal. Today 90 (2004) 183–190. Reproduced with kind permission of Elsevier.*

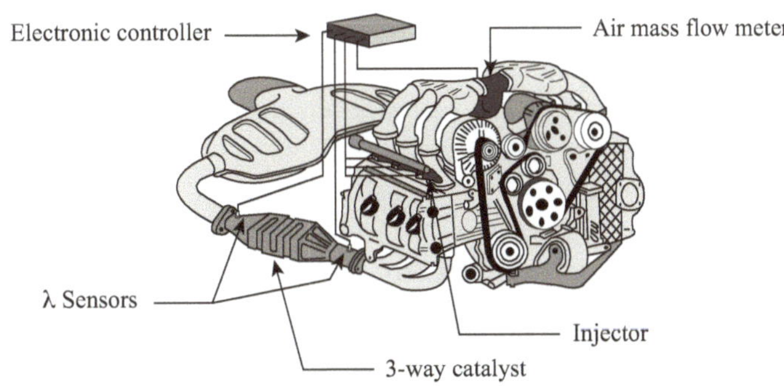

Figure 8.18:
Typical arrangement of the exhaust catalyst. *Source: J. Kasper et al., Catal. Today 77 (2003) 419–449. Reproduced with kind permission of Elsevier.*

feed to the engine. Figure 8.19 shows a schematic representation of the engine power and the concentrations of unburnt hydrocarbons, carbon monoxide and NO_x as a function of the air–fuel ratio; the position of the stoichiometric mix, that is the composition required to give 100% combustion of the fuel, is also shown by a vertical line. If there is an excess of air in the feed (left-hand side of the diagram), then the concentration of unburnt hydrocarbon increases and as does the amount of carbon monoxide formed. With an excess of air, the carbon monoxide concentration falls to zero and the unburnt hydrocarbon concentration also decreases. (The latter increases once more once the mixture enters the lean-burn region and gets worse still in the misfire region.) The formation of NO_x is relatively low in the fuel-rich region as a result of a lower temperature[26] and also of the

[26] NO_x is formed by a homogeneous reaction between oxygen and nitrogen in the engine cylinder. The reaction is temperature dependent.

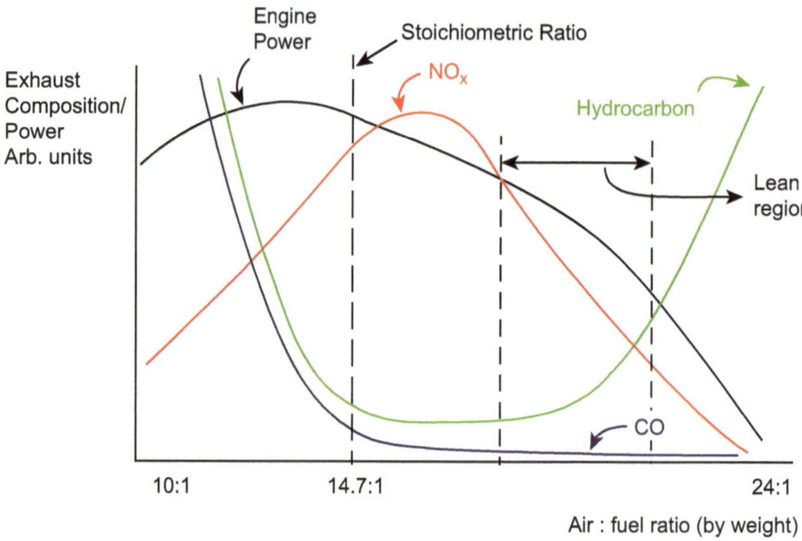

Figure 8.19:
The effect of the air–fuel ratio on engine operation.

relatively low concentration of oxygen present. It goes through a maximum at a gas composition close to the stoichiometric mixture and then drops once more when the mixture corresponds to the lean-burn region. The so-called three-way catalyst used for an engine operating close to the stoichiometric gas composition has, therefore, to lower the concentrations of NO_x, CO and unburnt hydrocarbon in an atmosphere containing largely N_2, CO_2 and water. Under these conditions, the concentration of oxygen is close to zero. Hence, the reactions occurring[27] are:

$$2CO + 2NO \rightarrow N_2 + 2CO_2$$

$$HC + NO \rightarrow CO_2 + H_2O + N_2$$

$$2H_2 + 2NO \rightarrow 2H_2O + N_2$$

(the hydrogen being produced by steam reforming of the unburnt hydrocarbon and also by the water–gas shift reaction). The original three-way catalysts contained largely Pt or a Pt/Rh mix supported on alumina but more recently Pd has also been used and an oxygen-storage component, a CeO_2–ZrO_2 mixed oxide, is also added. For more information on these catalyst compositions, the reader should see some of the recent reviews on the subject listed in Box 8.10.

[27] The reactions involving the hydrocarbon are not balanced.

With the introduction of more stringent requirements for fuel efficiency in the 1990s, the automobile manufacturers started to introduce lean-burn engines. These work with an excess of oxygen above the stoichiometric amount required for combustion of the fuel (see Figure 8.19). The three-way catalysts discussed above do not work effectively under these conditions. This led to a new area for research on selective NO_x reduction catalysts, notably in Japan, led by developments in Toyota Motor Corporation, first on selective de-NO_x using hydrocarbon reductants and then on NO_x storage catalysts. Some of the background is covered in the article by S. Matsamoto shown in Box 8.10 which also cites a number of important related academic publications up until that time.

BOX 8.10 Recent Reviews on Exhaust Catalysts and Reactions

The following reviews on the subject of exhaust gas clean-up provide useful entry points to the literature on this subject:

- V.I. Pârvulescu, P. Grange, B. Delmon, "Catalytic removal of NO", Catal. Today 46 (1998) 628.
- J. Kašpar, P. Fornasiero, N. Hickey, "Automotive catalytic converters: current status and some perspectives", Catal. Today 77 (2003) 419–449.
- K. Burch, J.P. Breen, F.C. Meunier, "A review of the selective reduction of NO_x with hydrocarbons under lean-burn conditions with non-zeolitic and platinum group metal catalysts", Appl. Catal. B (Environmental) 39 (2002) 283–303.
- S. Matsumoto, "Recent advances in automobile exhaust clean-up", Catal. Today 90 (2004) 183–190.
- W.S. Epling, L.E. Campbell, A. Yezerets, N.W. Currier, J.E. Parks II, "Overview of the fundamental reactions and degradation mechanisms in NO_x storage/reduction catalysts", Catal. Rev. Sci. Eng. 46 (2004) 163–245.
- M.V. Twigg, "Progress and future challenges in controlling automotive exhaust gas emissions", Appl. Catal., B (Environmental) 70 (2007) 2–15.

One of the most effective catalysts developed for use under lean-burn conditions is composed of silver supported on alumina. Work by Meunier et al. (F.C. Meunier, J.P. Breen, V. Zuzaniuk, M. Olsson, J.R.H. Ross, "Mechanistic aspects of the selective reduction of NO by propene over cobalt and silver promoted alumina catalysts: kinetic and in-situ DRIFTS study", J. Catal. 187 (1999) 493–505) showed that the reaction occurs on both silver crystallites and on the alumina surface and that $Ag+$ also participates, as shown schematically in Figure 8.20, and also that a variety of organic species containing nitrogen and oxygen were also formed. Unfortunately, although these catalysts give high conversions with high selectivities to nitrogen, they are particularly susceptible to impurities (particularly sulfur-containing molecules) in the fuels and so do not have the life-times required for practical use. As a result of this type of work, Toyota developed the so-called "NO_x storage catalysts", consisting of Pt and BaO supported on alumina, which are used under consecutively oxidizing and reducing conditions, adsorbing NO_x during the first oxidizing period and then reducing it during the reducing cycle.

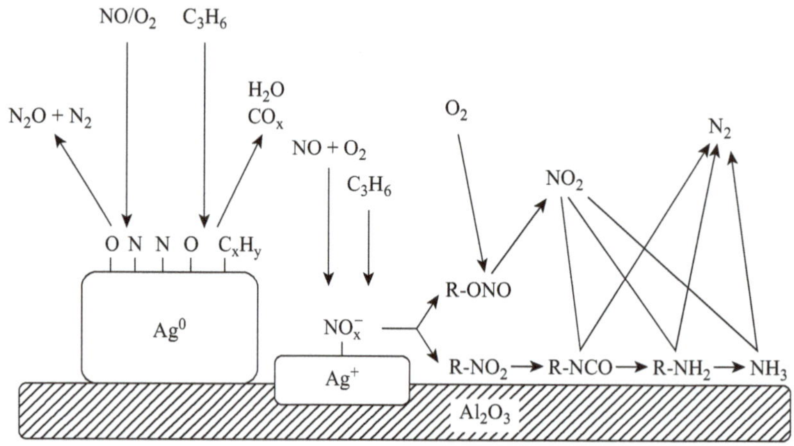

Figure 8.20:
Schematic representation of some of the steps occurring on a silver/alumina catalyst under lean-burn conditions. *Source: Meunier et al., J. Catal. 187 (1999) 493–505. Reproduced with kind permission of Elsevier.*

TASK 8.16 Lean-burn Engines

It should be stressed that the references given in Box 8.10 are not the only ones which could have been chosen. They have been selected as they are relatively highly cited in all cases and are also very relevant to the topics considered. Using the references of Box 8.10 as source, do a full literature search. List some of the catalyst types that have been studied for lean-burn conditions and for NO_x storage.

Figure 8.21 shows the model put forward by the Toyota researchers for the catalyst surface. It shows NO being oxidized on the Pt sites and being adsorbed as nitrate on Ba sites when the air–fuel ratio is lean; then, when the air–fuel ratio is changed to the stoichiometric value (reducing conditions), the nitrate is reduced by HC, CO and H_2O. A slight variant on this

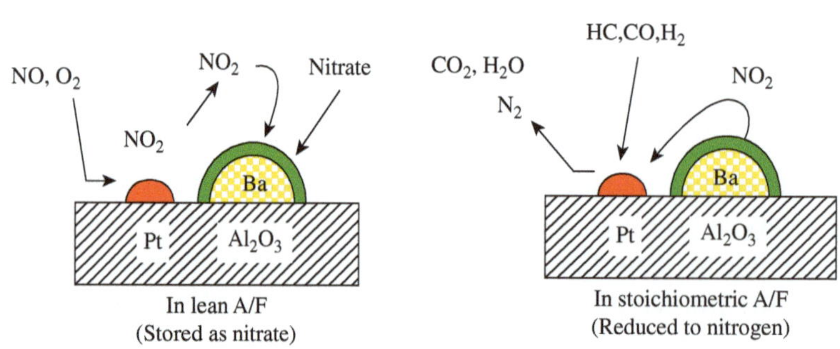

Figure 8.21:
The Toyota model for the NO_x storage catalyst. *Source: Matsumoto, Catal. Today 90 (2004) 183–190. Reproduced with kind permission of Elsevier.*

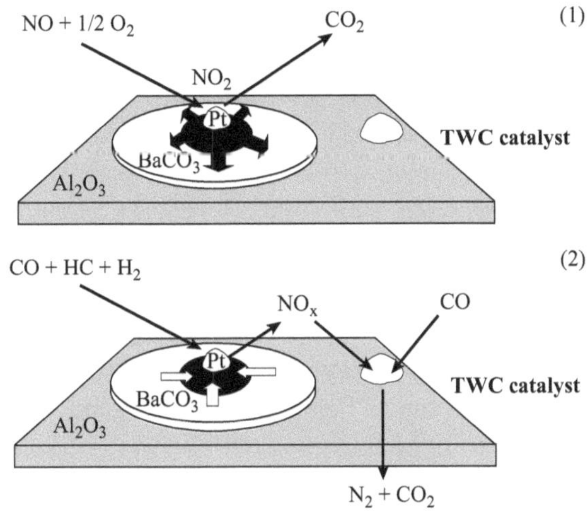

Figure 8.22:
Model for a NO_x storage catalyst including $BaCO_3$ formation. *Source: Kaspar et al., Catal. Today 77 (2003) 419–449. Reproduced with kind permission of Elsevier.*

picture is included in the review by Kaspar et al. (Box 8.10) as shown in Figure 8.22. In this model, it is recognized that the Ba present in the catalyst will form $BaCO_3$ and so the scheme shows its participation in the reaction.

There are many other types of environmental catalysts and processes which could also have been included in this chapter. It is worth pointing out that catalysis is even applied in a number of pieces of equipment used in the home: an example, taken from A. Nishino, "Household appliances using catalysis", Catal. Today, 10 (1991) 107–118, is shown in Figure 8.23.

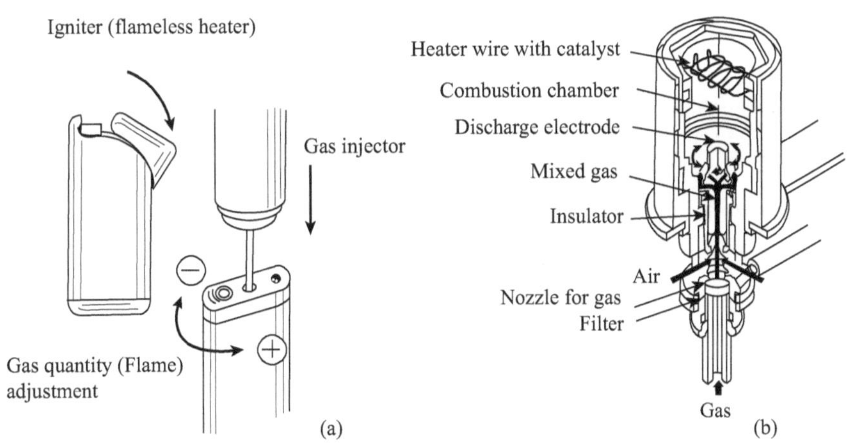

Figure 8.23:
A catalytic lighter. *Source: A. Nishino, Catal. Today 10 (1991) 107–118. Reproduced with kind permission of Elsevier.*

8.6 Catalysis in Biomass Conversion

This chapter would not be complete without some mention of a currently very important topic: biomass conversion. This topic has received a great deal of attention over the last few years as it is generally recognized that, with depleting oil reserves, the world has to seek other sources of, in particular, energy. While solar, wind, hydro and tidal energy are all feasible sources ready to be tapped, the engineering problems that need to be overcome for universal application are very considerable and they are not applicable in all situations. An alternative is nuclear energy, either from fission or from fusion. However, there are strong political pressures against the wider use of fission energy and fusion energy is still something of the future. Hence, the use of biomass has become of great importance. For example, in Europe, there is a mandatory target of 10% of energy requirements from biomass by 2020 and in the US the Energy Independence and Security Act of 2007 requires 36 billion gallons of renewable fuels per year by 2022.

In principle, biomass can be used for the production of both fuels and chemicals. Indeed, processes have existed for many years for the production of chemicals; a well-known example is the production of alcohol from plants containing sugar or starch such as potatoes, corn and sugar cane. Alcohol production from sugar cane has become a very important process in Brazil where a significant proportion of the requirements for motor fuel comes from ethanol produced by fermentation. Some proportion of Europe's energy supplies already comes from biomass but these are dependent on such technologies as biomass combustion (i.e. burning wood and other woody biomass instead of coal and oil) or from the use of so-called "first generation" processes using reactions such as the catalytic transesterification of triglycerides obtained from palm oil or from vegetable oils obtained from crops such as maize. These processes have come under attack because of the competition between food and energy that they represent. Further, it has been argued that the use of these crops can have an all-over negative effect on greenhouse gas production as the NO_x produced from the application of nitrate fertilizers will have a greater effect on global warming than the reduction in CO_2 emissions.

"Second generation biomass processes" have neither of the disadvantages of the first generation processes. They are based on the conversion of "lignocellulosic biomass" or, in other words, on biomass in which the main constituents are bound up in structures containing cellulose together with hemicellulose and lignin. The plants[28] used for such processes generally contain only small amounts of sucrose or the other molecules found in the first generation plants that are used largely as foodstuffs. Cellulose is a biopolymer made up of chains of sugar units which are so configured that there are strong hydrogen bonds between the chains as well as between the sugar rings (see Figure 8.24); this hydrogen bonding is not found

[28] Second generation feedstocks include specially grown crops such as miscanthus grass or switch grass but they may also be the residues from first generation production such as sugar bagasse or straw. Some second generation processes such as Biofine (see footnote 29) can also handle waste paper and municipal wastes.

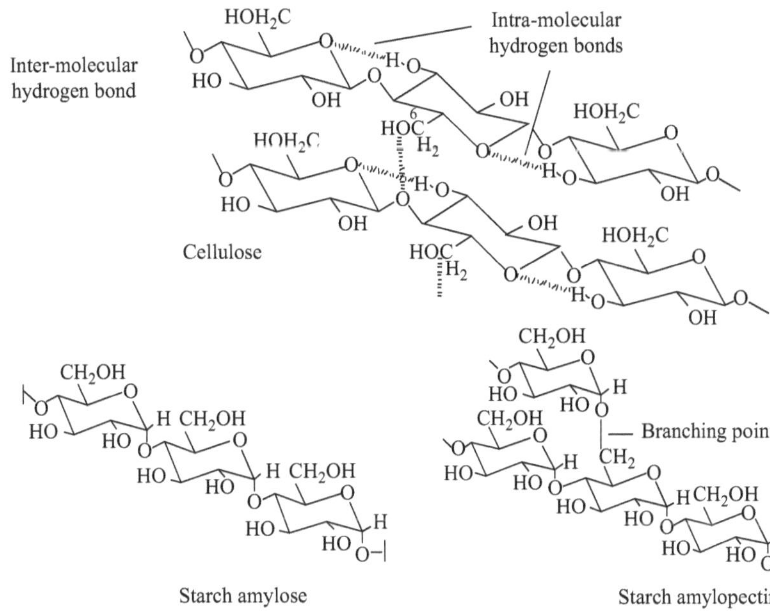

Figure 8.24:
The structures of cellulose and of the components of starch, amylose and amylopectin. *Source: D.J. Hayes, Catal. Today 145 (2009) 138–151. Reproduced with kind permission of Elsevier.*

in the starch structure which is made up of amylose and amylopectin units. (It should be noted that the amylose structure is similar to the building units in the cellulose but it lacks the β-configuration in the linkage between the sugar rings that is found in cellulose and it is therefore relatively easily hydrolyzed.) The cellulose structure is notoriously difficult to break down (depolymerize) and so the most critical part of any second generation plant is likely to be the preliminary cellulose treatment unit. Hayes gives details of a number of these pretreatment systems in the review paper listed in Box 8.11. As shown in the upper half of Figure 8.25, the cellulose fraction can then, on hydrolysis, give free monosaccharides which can be converted to ethanol or butanol or, with more complete hydrolysis (as in the so-called Biofine Process[29]), give levulinic acid, $CH_3COCH_2CH_2COOH$, plus formic acid. We will discuss some of the possible products from levulinic acid below. During the hydrolysis process, the lignin component of the lignocellulose, together with unconverted pentoses, form a "char" material. Gasification of this char (see the lower half of Figure 8.25 and below) is used to produce syngas which provides the energy needed to drive the process; if the scale is sufficiently high, the syngas can also be used for the production of chemicals.[30]

[29] J.J. Bozell, L. Moens, D.C. Elliott, Y. Wang, G.G. Neuenscwander, S.W. Fitzpatrick, R.J. Bilski, J.L. Jarnefeld, "Production of levulinic acid and use as a platform chemical for derived products", Res. Conserv. Recycl., 28 (2000) 227–239.

[30] After the gasification process, there generally still remains a solid residue which in Figure 8.25 is denoted as Biochar; this has been used successfully as a soil enhancer.

BOX 8.11 Reviews on Biomass Conversion

A useful review on biomass usage and the relevant processes is given by D.J. Hayes, "An examination of biorefining processes, catalysts and challenges", Catal. Today 145 (2009) 138–151. The review by G.W. Huber et al. covers many aspects of the background to biomass conversion as well as giving an excellent overview of biomass conversion processes: G.W. Huber, S. Iborra, A. Corma, "Synthesis of transport fuels from biomass: chemistry, catalysts and engineering", Chem. Rev. 106 (2006) 4044–4098. A further review by the same group gives additional material: A. Corma Canos, S. Iborra, A. Velty, "Chemical routes for the transformation of biomass into chemicals", Chem. Rev. 107 (2007) 2411–2502. While the two reviews in *Chemical Review* will not be accessible to you through Science Direct, the lists of references provided and the many citations listed in the Scopus or Web of Science entries for these reviews will give an excellent coverage of the whole area of biomass conversion.

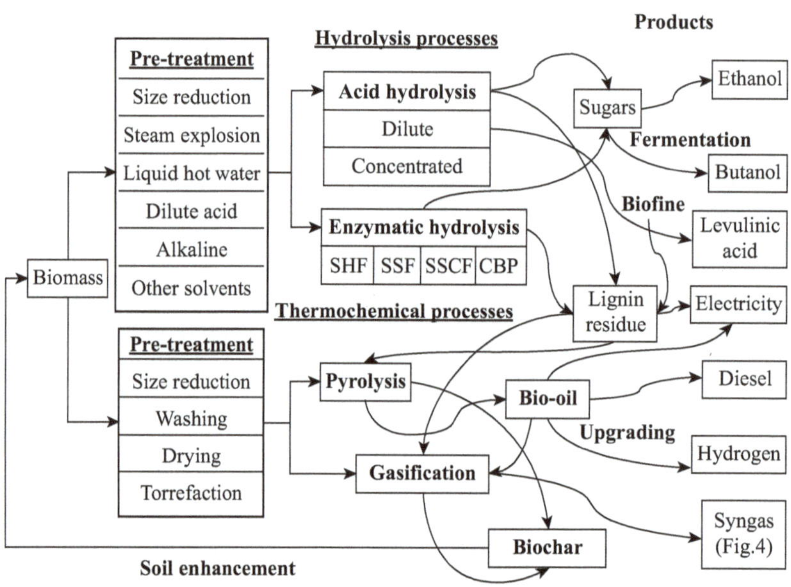

Figure 8.25:
Lignocellulosic biomass conversion to useful products. *Source: D.J. Hayes, Catal. Today 145 (2009) 138–151. Reproduced with kind permission of Elsevier.*

Direct gasification of biomass is also a well-used option. Figure 8.26 shows a schematic representation of the options available for gasification technology; this figure is taken from a comprehensive review on the subject by G.J. Stiegel and R.C. Maxwell, "Gasification technologies: the path to clean, affordable energy in the 21st century", Fuel Process. Technol. 71 (2001) 79–97.[31] Catalysis has a significant role to play in gas clean-up, the shift reaction

[31] Several of the citations to this article in Scopus refer directly to biomass gasification.

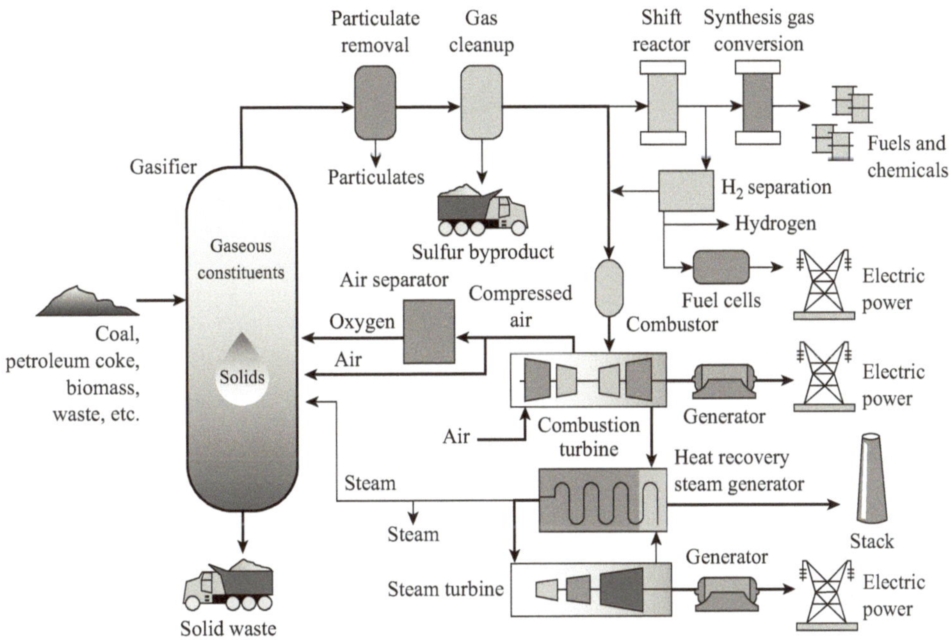

Figure 8.26:
Gasification technology options. *Taken with permission from G.J. Stiegel, R.C. Maxwell, Fuel Process. Technol. 71 (2001) 79–97. Reproduced with kind permission of Elsevier.*

and synthesis gas conversion. We have already discussed the shift reaction and synthesis gas conversion, technologies that are applied elsewhere. Gas clean-up is more specific to gasification technology and generally involves removal of gases such as methane and other hydrocarbon molecules from the gas stream as well as the destruction of tars, etc. The catalysts often used are related to those used for steam reforming. The subject of gas clean-up has been reviewed by Sutton et al. (see Task 8.17).

Once levulinic acid has been produced by the Biofine Process, it can be converted to wide range of products using catalysis. Figure 8.27 shows one such route to produce

TASK 8.17 Gas Clean-up

The subject of the clean-up of the syngas produced by gasification has been reviewed by D. Sutton, B. Kelleher, J.R.H. Ross, "Review of literature on catalysts for biomass gasification", Fuel Process. Technol., 73 (2001) 155–173. Another paper that reports a reactor system for dealing with the conversion of the tars produced is the following: P. Ammendiola, B. Piriou, L. Lisi, G. Ruoppolo, R. Chicore, G. Russo, 'bed reactor for the study of catalytic biomass tars conversion', Experimental Thermal Fluid Sci. 34 (2010) 264–274. Use these papers as sources from which to examine the earlier literature on this subject and, using Scopus or Web of Science, the relevant literature citing it.

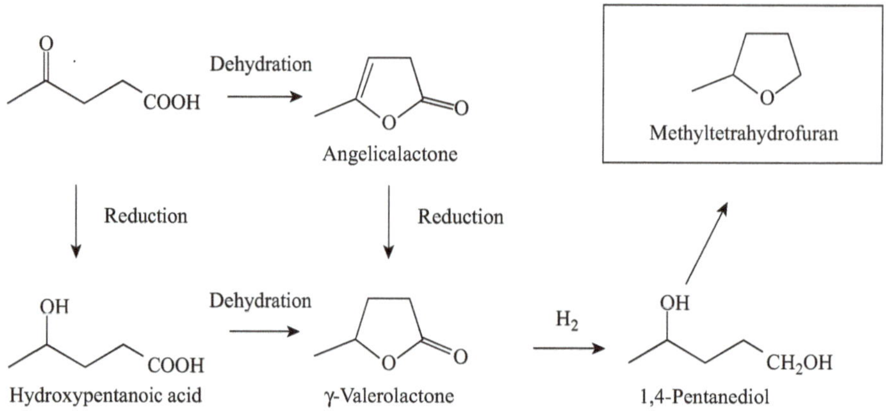

Figure 8.27:
Catalytic conversion of levulinic acid produced in the Biofine Process to various products. *Source: J. J., Bozell, L. Moens, D. C. Elliott, Y. Wang , G. G. Neuenscwander, S. W. Fitzpatrick, R. J. Bilski J. L. Jarnefeld, "Production of levulinic acid and use as a platform chemical for derived products", Res. Conserv. Recycl. 28 (2000) 227–239. Reproduced with kind permission of Elsevier.*

methyltetrahydrofuran, a molecule that can be used as a petroleum additive at concentrations up to 30%. A range of other products are also discussed in the paper by Bozell et al. Shell has recently shown that it is also possible to make so-called "valeric biofuels" from levulinic acid.[32]

TASK 8.18 Catalysis in the Conversion of Levulinic Acid

Use the papers by Bozell et al. (see Figure 8.27) and Hayes (see Figure 8.24) as sources for citation searches on the conversion of levulinic acid to valuable products. Include in your material the results shown in the paper of footnote 32 and also search for papers by L.E. Manzer.

A rival technology to those involving hydrolysis or direct gasification is one making use of biomass pyrolysis. Depending on the conditions used (temperature, pressure and residence time) different proportions of pyrolysis oil, gas and char are produced. An excellent review of the effects of different conditions has been published by A.V. Bridgwater, "Renewable fuels and chemicals by thermal processing of biomass", Chem. Eng. J. 91 (2003) 87–102. In this, he pays particular attention to the operation of the fast pyrolysis process which produces, at several seconds residence time and at a temperature of about 500 °C, a mixture of 75% liquids, 12% char and 13% gas. The liquid is a relatively unstable micro-emulsion that contains about 25% water and is made up of a mixture of oxygenates (particularly acetic acid), a solid char and

[32] Shell has recently shown that it is possible to make so-called valeric biofuels from levulinic acid: J.-P. Lange, R. Price, P.M. Ayoub, J. Louis, L. Petrus, L. Clarke, H. Gosselink, "Valeric biofuels: A platform of cellulosic biofuels", Angew. Chem., Int. Ed., 49 (2010) 4479–4483.

alkali metals originating from the ash present. In his review, Bridgwater discusses the nature of the bio-oil and some of the products which can be obtained from it, particularly by hydrotreating or by cracking over zeolites. In the context of this chapter, it is probably of greatest relevance to note that significant amounts of work have been carried out on the upgrading of acetic acid derived from bio-oil. An example of such work is included in Task 8.19.

TASK 8.19 Steam Reforming of Acetic Acid from Bio-Oil

A well-cited paper on steam reforming of acetic acid obtained from bio-oil is the following: T. Takanabe, K.I. Aika, K. Seshan, L. Lefferts, "Sustainable hydrogen from bio-oil: Steam reforming of acetic acid as a model oxygenate", J. Catal. 227 (2004) 101–108. Use this article as an entrance point to the literature on this subject, examining the types of catalysts used for such work and the best results obtained for hydrogen selectivities and yields.

8.7 Conclusions

This chapter has had as its aim to cover some, but by no means all, of the applications of catalysis in practice which have received a lot of attention over the last few decades. Of necessity, these have in many cases been applications of which the present author has had some experience. No attempt has been made here, or indeed elsewhere in the book, to provide detailed listings of relevant literature. It is hoped instead that the reader has followed up in some detail, using any web-based facilities at their disposal a sufficiently large number of distinct topics in order that he or she has gained a significant amount of new knowledge and has been convinced of the author's belief that it is not what you know but knowing how to find it that is most important.

Index

Kolibri 360 GmbH
Ferdinand-Jühlke-Straße 7
99095 Erfurt, Deutschland
produktsicherheit@kolibri360.de